Photographic and Descriptive Musculoskeletal

Atlas of Gibbons and Siamangs (*Hylobates*)

With notes on the attachments, variations, innervation, synonymy and weight of the muscles

Photographic and Descriptive Musculoskeletal Atlas of Gibbons and Siamangs (*Hylobates*)

With notes on the attachments, variations, innervation, synonymy and weight of the muscles

- Rui Diogo
- Juan F. Pastor
- Eva M. Ferrero
- Mercedes Barbosa
- Anne M. Burrows
- Bernard A. Wood

- Josep M. Potau
- Félix J. de Paz
- Gaëlle Bello
- M. Ashraf Aziz
- Julia Arias-Martorell

CRC Press
Taylor & Francis Group
Boca Raton London New York

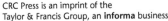
CRC Press is an imprint of the
Taylor & Francis Group, an **informa** business

Science Publishers
Jersey, British Isles
Enfield, New Hampshire

CRC Press
Taylor & Francis Group
6000 Broken Sound Parkway NW, Suite 300
Boca Raton, FL 33487-2742

First issued in paperback 2019

© 2012 by Taylor & Francis Group, LLC
CRC Press is an imprint of Taylor & Francis Group, an Informa business

No claim to original U.S. Government works

ISBN-13: 978-1-57808-786-0 (hbk)
ISBN-13: 978-0-367-38150-9 (pbk)

CIP data will be provided on request.

**Visit the Taylor & Francis Web site at
http://www.taylorandfrancis.com**

**and the CRC Press Web site at
http://www.crcpress.com**

Acknowledgements

We gratefully acknowledge support and funding from all of the institutions and funding bodies that made this project possible; we especially acknowledge GW's support for RD's research via a GW Presidential Fellowship. We particularly acknowledge Brian Richmond (George Washington University) for allowing us to dissect the GWU HL1 specimen, Gonzalo Fernández and Jesús Recuero (Bioparc Fuengirola) for providing the VU HG1 and VU HG2 specimens, and Chris Bonar (Dallas World Aquarium and Zoological Garden of Dallas) for procuring the DU HM1 and DU HS1 specimens when he was at the Cleveland MetroParks Zoo.

Contents

CHAPTER 1

Introduction and Aims

Gibbons and siamangs (*Hylobates*), apart from common chimpanzees and bonobos, gorillas and orangutans, are our closest living relatives. This photographic and descriptive musculoskeletal atlas of *Hylobates* follows the same configuration used in the photographic atlas of *Gorilla* published in 2010 (Diogo et al. 2010). These two books are part of a monograph series that will also include atlases of orangutans and chimpanzees. The series is designed to provide the comparative, phylogenetic, and evolutionary context for understanding the evolutionary history of the gross anatomy of modern humans and our closest relatives.

We dissected and took high-quality photographs of various specimens from different species of hylobatids, including infants, juvelines and adults and both males and females (see Methodology and Material below) and for one of these specimens (VU HG2) we were able to record the wet weight of many of the muscles. Where there are differences between the myology (e.g., the presence/absence of a muscle or a muscle bundle, its attachments and/or its innervation) of this specimen and that of the other specimens dissected by us, we provide detailed comparative notes, and use photographs to document the differences.

The atlas also includes the results of an extensive review of the literature about the musculature of gibbons and siamangs, a comprehensive review of muscle variants among individual hylobatids, and a list of the synonyms used in the literature to refer to the muscles of these primates. The data previously obtained from our dissections of numerous primates and other mammalian and non-mammalian vertebrates (e.g., Diogo 2004a,b, 2007, 2008, 2009; Diogo and Abdala 2007, 2010; Diogo et al. 2008, 2009a,b, 2010; Diogo and Wood 2011a,b) were used to test hypotheses about the homologies among the muscles of hylobatids, great apes and modern humans, and other taxa.

We hope this atlas will be of interest to students, teachers and researchers studying primatology, comparative anatomy, functional morphology, zoology, and physical anthropology, as well as to clinicians and researchers who are interested in understanding the origin, evolution, and homology of the musculoskeletal system of modern humans as well as the comparative context of common variants on the musculature of modern humans.

CHAPTER 2

Methodology and Material

The hylobatid specimens dissected for this study were made available by the following institutions: George Washington University (*H. lar* specimen GWU HL1, juvenile female), Howard University (*H. lar* specimen HU HL1, adult male; this specimen came from the Yerkes Regional Primate Center where it had the number YN87-134), Duquesne University (*H. muelleri* specimen DU HM1, adult male; *H. syndactylus* specimen DU HS1, adult male; these two specimens were provided by the Cleveland MetroParks Zoo), and Valladolid University (*H. gabriellae* specimen VU HG1, 1-year old infant male; *H. gabriellae* specimen VU HG2, adult male; *H. klossii* specimen VU HK1, adult male; these three specimens were provided by the Bioparc Fuengirola). It should be noted that some authors recognize two extant genera of hylobatids: *Hylobates* (to include the "gibbons") and *Symphalangus* (to include the "siamangs") (e.g., Goodman 1999). However, recent genetic studies indicate that "gibbons" are not a monophyletic group in that most "gibbons" are apparently more closely related to siamangs than to the "gibbons" of the subgenus *Nomascus* (Fabre et al. 2009; Arnold et al. 2010). Other authors (including Fabre et al. 2009) recognize four extant hylobatid genera but they do not agree on the identity of those genera. For example, some refer to *Hylobates*, *Bunopithecus*, *Symphalangus* and *Nomascus*, while others refer to *Hylobates*, *Hoolock*, *Symphalangus* and *Nomascus* (see Groves 2001, 2005). We use a more traditional and stable (e.g., Napier and Napier 1985; Nowak 1999; Groves 2001) classification that recognizes a single extant hylobatid genus (*Hylobates*; including species such as *H. syndactylus*, *H. lar*, *H. gabriellae* and *H. hoolock*, among others). Readers should therefore be aware that the vernacular term "gibbons" used in this atlas and in other publications about hylobatids almost certainly refers to a group that is not monophyletic.

We took photographs of the musculoskeletal system of all dissected hylobatid specimens, but the muscle weights listed in this atlas are from the *H. gabriellae* specimen VU HG2 (total body weight = 5.200 kg), which was a fresh specimen that was in particularly good condition. The photographs of the osteological structures shown in this atlas are from the *H. gabriellae* specimen HU HG2 and the *H. klossii* specimen VU HK1. In the text below, we provide for each muscle: 1) its weight in the *H. gabriellae* VU HG2 specimen (it is not possible to accurately measure the mass of all the muscles, but when it can be measured the value we obtained is given in parentheses immediately following the name of the muscle; when the muscle is paired, e.g., stylohyoideus, the weight given is that of the muscle of one side of the body; when the muscle is unpaired, e.g., arytenoideus transversus, the weight given is that of half of the muscle, i.e., also from a single side of the body); 2) the most common attachments and innervation of the muscle within hylobatids, based on our dissections and on our literature review; 3) comparative notes, for instance in those cases where there are differences (e.g., regarding the presence/absence of the muscle, or of its bundles, its attachments, and/or its innervation) between the configuration usually found in hylobatids and the configuration found in a specimen dissected by us (in these cases we often provide photographs to illustrate the differences) or by others; 4) a list of the synonyms that have been used by other authors to designate that muscle.

Apart from the hylobatid specimens mentioned above, we have dissected numerous specimens from most vertebrate groups, including bony fish, amphibians, reptiles, monotremes, rodents, colugos, tree-shrews, and numerous primates, including modern humans (a complete list of these specimens and the terminology used to describe them is given in Diogo and Abdala 2010 and Diogo and Wood 2011a). This broad comparative context proved to be crucial for generating hypotheses about the homologies among the muscular structures of hylobatids, modern humans and other primate and non-primate vertebrates, and it also informed the nomenclature proposed by Diogo et al. (2008, 2009a,b), Diogo and Abdala (2010) and Diogo and Wood (2011a). This nomenclature is based on that employed in modern human anatomy (e.g., Terminologia Anatomica 1998), but it also takes into account the names used by researchers who have focused on non-human mammals (e.g., Saban 1968; Jouffroy 1971). In the majority of the figures we use Latin names for the soft tissues and anglicized names for the bones. In the figures that mainly illustrate osteological structures we use Latin names, but to avoid redundancy when these names are similar to the anglicized version (e.g., processus mastoideus = mastoid process) we do not provide the latter; in those cases in which they are substantially different (e.g., incisura mandibulae and mandibular notch) we provide both the Latin names and the anglicized version.

When we describe the position, attachments and orientation of the muscles and we use the terms anterior, posterior, dorsal and ventral in the sense in which those terms are applied to pronograde tetrapods (e.g., the sternohyoideus mainly runs from the sternum, posteriorly, to the hyoid bone, anteriorly, and passes mainly ventrally to the larynx, which is, in turn, ventral to the esophagus; the flexors of the forearm are mainly situated on the ventral side of the forearm). However, the nomenclature used in Terminologia Anatomica (1998) was defined on the basis of an upright posture and although most primates are not bipeds, nearly all of the osteological names and most of the myological ones used by other authors (and by us) to designate the structures of non-human primates, including hylobatids, follow the Terminologia Anatomica nomenclature. Although this is potentially confusing we judged it to be preferable to refer to the topology of the musculoskeletal structures of non-human primates in this way because in the vast majority of primates the 'superior angle of the scapula' is actually mainly anterior, and not superior, to the 'inferior angle of the scapula', and the 'cricoarytenoideus posterior' actually lies more on the dorsal, and not on the posterior, surface of the larynx. Moreover, we think that, by keeping in mind that the actual names (both in Latin and in English) of all the osteological structures and of most the myological structures mentioned in this atlas refer to a biped posture while the actual descriptions provided here regarding the topology of these structures refer to a pronograde posture, most readers will have no difficulty in interpreting and understanding the information provided in this book.

The muscles listed below are those that are usually present in adult hylobatids; muscles that are only occasionally present in adult hylobatids are discussed in other parts of the atlas. In our written descriptions, we follow Edgeworth (1935), Diogo and Abdala (2010) and Diogo and Wood (2011a,b) and divide the head and neck muscles in five main subgroups: 1) mandibular, muscles that are generally innervated by the fifth cranial nerve (CN5) and include the masticatory muscles, among others; 2) hyoid, muscles that are usually innervated by CN7 and include the facial muscles, among others; 3) branchial, muscles that are usually innervated by CNC9 and CN10, and include most laryngeal and pharyngeal muscles, among others; 4) hypobranchial, muscles that include all the infrahyoid and tongue muscles, and the geniohyoideus. According to Edgeworth (1935) the hypobranchial muscles were developed primarily from the anterior myotomes of the body and then migrated into the head. Although they retain a main innervation from spinal nerves, they may also be innervated by CN11 and CN12, but they usually do not receive any branches from CN5, CN6, CN7, CN8, CN9 and CN10; 5) extra-ocular, muscles that are usually innervated by nerves CN3, CN4 and/or CN6 in vertebrates. The head, neck, pectoral and upper limb muscles are listed following the order used by Diogo et al. (2008, 2009a), Diogo and Abdala (2010) and Diogo and Wood (2011a,b), while the pelvic and lower limb muscles, as well as the other muscles of the body, are listed following the order used by Gibbs (1999). It should be emphasized that the literature review undertaken by this latter author provided a crucial basis and contribution for our own literature review.

CHAPTER 3

Head and Neck Musculature

3.1 Mandibular musculature

Mylohyoideus (Fig. 8)
- Usual attachments: From the mylohyoid line of the mandible to the hyoid bone, posteriorly, and to the ventral midline, anteriorly.
- Usual innervation: Mylohyoid nerve of mandibular division of CN5 (Kohlbrügge 1890–1892: *H. syndactylus, H. agilis*).
- Notes: There is usually no distinct median raphe of the mylohyoideus in hylobatids, according to Saban (1968). However, Dubrul (1958) did describe a median raphe in *H. lar*. Wall et al. (1994) stated that stimulation of the mylohyoideus in *H. lar* elicited slight hyoid elevation and tongue protrusion.
- Synonymy: Intermandibularis (Edgeworth 1935).

Digastricus anterior (1.4 g; Figs. 5, 8)
- Usual attachments: From the intermediate tendon of the digastric, and sometimes also from the hyoid bone, to the mandible.
- Usual innervation: Mylohyoid nerve of mandibular division of CN5 (Kohlbrügge 1890–1892: *H. syndactylus, H. agilis*).
- Notes: Some contact seems to be present between the digastricus anterior and its counterpart (at the midline) in the *Hylobates* specimen shown in Fig. 1 of Wall et al. (1994), and Gibbs et al. (2002) suggested that this is the usual condition in hylobatids. However, according to the detailed studies of Deniker (1885), Kohlbrügge (1890–1892) and Dubrul (1958) and to our dissections, the usual condition for hylobatids is that there is no contact for most of the length of the muscle. According to Dubrul (1958) in *H. lar* the fleshy parts of the anterior digastric muscles are separated, but the posterior portion of these structures as well as the tendons connecting them to the posterior digastric muscles are related to "complicated fascia plates" that attach to the hyoid bone and to their counterparts at the midline; anteriorly the anterior digastric muscles attach onto the lower border and inner surface of the mandible near to the symphysis. Wall et al. (1994) stated that in *H. lar* the posterior portion of the anterior

digastrics attaches to the intermediate tendons of the posterior digastrics and aponeurotically to the body of the hyoid. The intermediate tendons join to form a tendinous arch situated between the mandibular symphysis and the anterior border of the hyoid bone; laterally, these intermediate tendons are situated cranial to the junction of the body and greater horn of the hyoid and are held in place at this junction by fascia. According to Wall et al. (1994) stimulation of the digastricus anterior in *H. lar* elicited mandibular depression and anterocranial displacement of the hyoid bone.
- Synonymy: Part of biventer maxillae superioris (Kohlbrügge 1890–1892).

Tensor tympani
- Usual attachments: From the auditory tube and adjacent regions of the neurocranium to the manubrium of the malleus.
- Usual innervation: Data are not available.
- Notes: According to Maier (2008) in hylobatids the chorda tympani usually passes above the tensor tympani (i.e., the epitensoric condition).

Tensor veli palatini
- Usual attachments: From the Eustachian tube and the adjacent regions of the neurocranium to the pterygoid hamulus and soft palate.
- Usual innervation: Data are not available.
- Notes: In modern human infants and in adult apes, including hylobatids, the palate lies much closer to the roof of the nasopharynx than it usually does in adult modern humans, so in the former the levator veli palatini and tensor veli palatini do not run so markedly downwards to reach the palate as they do in the latter.

Masseter (Figs. 5, 6, 8, 9, 11, 12, 13)
- Usual attachments: Mainly from the zygomatic arch; the pars superficialis inserts mainly onto the lower edge of the base of the mandible, while the pars profunda inserts mainly onto the ascending ramus and the coronoid process of the mandible.
- Usual innervation: Mandibular division of CN5 (Kohlbrügge 1890–1892: *H. syndactylus, H. agilis, H. moloch*).
- Notes: The mean size ratio between the origin and insertion of the superficial masseter is 1.29 in *Hylobates* (in contrast to 0.99 in *Homo* and 0.78 in *Pongo*: Gibbs 1999). In the hylobatids dissected by us the zygomatico-mandibularis does not seem to be present as a distinct structure, and there is seemingly no strong tissue between the two heads of the masseter; some of the fibers of the superficial head are blended with fibers of the pterygoideus medialis. Wall et al. (1994) stated that stimulation of the superficial head of the masseter in *H. lar* resulted on mandibular elevation and protraction.

Temporalis (Figs. 6, 8, 9, 11, 12)

- Usual attachments: From the whole of the fossa temporalis, the temporalis fascia and the adjacent region of the skull to the coronoid process and the ascending ramus of the mandible.
- Usual innervation: Mandibular division of CN5 (Kohlbrügge 1890–1892: *H. syndactylus, H. agilis, H. moloch*).
- Notes: Edgeworth (1935) stated that in *Hylobates* the temporalis is inserted through a separate tendon onto the junction of the coronoid process and the body of the mandible. In the specimens dissected by us (e.g., in our *H. gabriellae* specimen VU HG1) the temporalis does not have a distinct pars suprazygomatica and it is seemingly not differentiated into a pars profunda and a pars superficialis. A pars suprazygomatica was not described by other authors (e.g., Kohlbrügge 1890–1892) in hylobatids.

Pterygoideus lateralis (Fig. 6)

- Usual attachments: From the lateral pterygoid plate and the adjacent regions of neurocranium to the capsule of temporomandibular joint and the neck of the mandibular condyle.
- Usual innervation: Mandibular division of CN5 (Kohlbrügge 1890–1892: *H. syndactylus, H. agilis, H. moloch*).
- Notes: In his description of *Hylobates* Kohlbrügge (1890–1892) did not refer to a differentiation of the pterygoideus lateralis into a caput inferius and a caput superius, but these two bundles were clearly present in the hylobatids dissected by us (e.g., in our *H. gabriellae* specimen VU GG1) and by others (e.g., Wall et al. 1994). Wall et al. (1994) stated that stimulation of the inferior head in *H. lar* resulted in pronounced mandibular depression.
- Synonymy: Pterygoideus externus (Kohlbrügge 1890–1892).

Pterygoideus medialis (Figs. 6, 13)

- Usual attachments: Mainly from the medial lamina of the pterygoid process and from the pterygoid fossa to the medial side of the mandible.
- Usual innervation: Mandibular division of CN5 (Kohlbrügge 1890–1892: *H. syndactylus, H. agilis, H. moloch*).
- Synonymy: Pterygoideus internus (Kohlbrügge 1890–1892).

3.2 Hyoid musculature

Stylohyoideus (Fig. 8)

- Usual attachments: From the tympanic region (the styloid process is usually small or absent in hylobatids) to the hyoid bone.
- Usual innervation: CN7 (Kohlbrügge 1890–1892: *H. syndactylus, H. agilis, H. moloch*).

- Notes: The usual condition for hylobatids is no piercing of the stylohyoideus by the digastricus posterior and/or the intermediate digastric tendon; there is no piercing in the fetal *Hylobates* sp. specimen dissected by Deniker (1885) and in 5 of the 6 sides of the 3 specimens reported by Kohlbrügge (1890–1892), although there was a partial piercing on the right side of one of these three specimens, and in the specimen reported by Bischoff (1870). We could not check this feature in the hylobatid specimens dissected by us. Clegg (2001) stated that in the *H. muelleri* specimen dissected by her the stylohyoideus originated "from a small styloid process, although this may have been ossified muscle", Deniker (1885) reported that in the fetal *Hylobates* sp. specimen dissected by him the stylohyoideus originated from the 'cartilage stylien' (near the origin of the styloglossus), and Kohlbrügge (1890–1892) stated that in two *Hylobates syndactylus* specimens and one *Hylobates agilis* specimen dissected by him the muscle originated from the tympanic region (there was no styloid process). To our knowledge the **jugulohyoideus** (usually present in strepsirrhines and sometimes present in *Tarsius*) and the **stylolaryngeus** (sometimes present in orangutans) have never been reported in hylobatids.

Digastricus posterior (0.8 g; Figs. 5, 8)
- Usual attachments: From the mastoid region and the adjacent part of temporal bone to the intermediate tendon.
- Usual innervation: CN7 (Kohlbrügge 1890–1892: *H. syndactylus, H. agilis, H. moloch*).
- Notes: Wall et al. (1994) stated that stimulation of the digastricus posterior in *H. lar* resulted in depression and slight retraction of the mandible. Contrary to the actions of the digastricus anterior, stimulation of the digastricus posterior did not result in the anterocranial displacement of the hyoid bone; this supports the idea that the digastricus posterior has no direct attachment onto the hyoid bone. Kohlbrügge (1890–1892) did not report any direct attachment onto the hyoid bone in the two *Hylobates syndactylus* specimens and the *Hylobates agilis* specimen dissected by him, and we also did not find such an attachment in the specimens dissected by us (e.g., in our *H. gabriellae* specimen VU HG1). However, Deniker (1885) stated that in the fetal *Hylobates* sp. specimen dissected by him the digastricus posterior inserted onto the hyoid bone, and Clegg (2001) reported a *H. muelleri* specimen in which the muscle attached to the hyoid bone by what appeared to be a tendon.

Stapedius
- Usual attachments: Probably inserts onto the stapes, but the information provided in the literature is sparse.
- Usual innervation: Data are not available.

Platysma cervicale (Fig. 1)
- Usual attachments: Mainly from the nuchal region to the modiolus and to adjacent regions of the mouth.

- Usual innervation: Branches of CN7.
- Notes: Influential authors such as Owen (1830-1831) and Sonntag (1924) used the name 'platysma myoides' to describe the platysma complex of Asian apes such as orangutans, and this nomenclature has been followed by various researchers, including Seiler (1976), and was thus also followed in Diogo et al.'s (2009b) review. However, Owen (1830-1831) stated that the 'platysma myoides' of orangutans incorporates the platysma myoides of modern humans plus the platysma cervicale of other mammals, and our recent dissections of numerous primates and comparisons with the data provided in the literature confirm that orangutans and hylobatids usually have a platysma cervicale, i.e., a part of the platysma complex that attaches onto the nuchal region. That is, juvenile and adult orangutans and hylobatids usually have a well-developed platysma cervicale similar to the muscle that is found in most other primates and that is usually markedly reduced, or even absent, in *Pan* (including the neonates dissected by us), *Gorilla* and *Homo*. Deniker (1885) described a single 'platysma' in the fetal *Hylobates* sp. specimen dissected by him, but stated that posteriorly this 'platysma' had various bundles and passed just inferiorly to the ear to cover most of the nuchal region almost making contact with its counterpart at the dorsal midline; the structure corresponding to the platysma myoides *sensu* the present study decussates with its counterpart at the ventral midline just posterior to the mandibular symphysis. In Fig. 33 of Huber (1930b) and Fig. 8 of Huber (1931), this author shows a *H. pileatus* specimen with a 'deteriorating nuchal platysma' (platysma cervicale *sensu* the present study) and a well-developed platysma myoides. The gibbon and siamang infant and adult specimens reported by Seiler (1976; e.g., *H. lar, H. agilis, H. moloch, H. syndactylus*) clearly have a platysma cervicale going to the nuchal region (see e.g., his Figs. 132 and 135). In our *H. lar* specimen HU HL1 and our *H. gabriellae* specimen VU HG1 the platysma cervicale is clearly present, reaching the dorsal midline in a significant portion of the dorsal margin of the neck, so it is not a 'deteriorating' structure as stated by Huber (1930b, 1931), but instead a well-developed muscle. Anteriorly the platysma cervicale is deeply blended with the platysma myoides (which mainly originates from the shoulder region) and extends anteriorly, passing mainly deep to the depressor anguli oris, except for a few of its anteroventromedial fibers that pass medial to this latter muscle. Anteriorly the platysma cervicale plus platysma myoides were partially inserted onto the mandible and partially blend with fibers of the depressor anguli oris, of the levator anguli oris facialis, of the orbicularis oris, and of the depressor labii inferioris.

Platysma myoides (Figs. 1, 3, 4)
- Usual attachments: Main from the pectoral region and the neck to the modiolus and adjacent regions of the mouth.
- Usual innervation: Branches of CN7.

Head and Neck Musculature   11

- Notes: See platysma cervicale above. The **sphincter colli superficialis** and **sphincter colli profundus** are not present as distinct muscles in hylobatids.

Occipitalis (Fig. 4)
- Usual attachments: From the occipital region to the galea aponeurotica and the ear region.
- Usual innervation: Branches of CN7.
- Notes: According to Deniker (1885), Ruge (1911), Huber (1930b, 1931), Loth (1931), Edgeworth (1935), Seiler (1976) and to our dissections in hylobatids the occipitalis is usually differentiated into a main body (or 'occipitalis proprius') and a 'cervico-auriculo-occipitalis' (*sensu* Lightoller 1925, 1928, 1934, 1939, 1940a,b, 1942, which is a lateral/superficial bundle of the occipitalis that often runs anterolaterally from the occipital region to the posterior portion of the ear and that sometimes covers part of the auricularis posterior in lateral view). Deniker (1885) stated that on the right side of the fetal *Hylobates* sp. specimen dissected by him the 'auriculaire postérieur' was 'simple', but that on the right side it had two bundles; one of these two bundles clearly seems to correspond to the 'cervico- auriculo-occipitalis' *sensu* the present study.
- Synonymy: Occipitalis plus part of the auricularis posterior (Deniker 1885, Edgeworth 1935).

Auricularis posterior (Fig. 1)
- Usual attachments: From the occipital region to the posterior region of the ear.
- Usual innervation: Branches of CN7.
- Notes: See occipitalis above.
- Synonymy: Part of the auricularis posterior (Deniker 1885, Edgeworth 1935).

Intrinsic facial muscles of ear (Fig. 4)
- Usual attachments: See notes below.
- Usual innervation: Branches of CN7.
- Notes: The intrinsic facial muscles of the ear of hylobatids were almost never described in the literature, and were difficult to analyze in our specimens, but Ruge (1911) and Seiler (1976) did examine these muscles in some detail. According to Seiler (1976) the **helicis, antitragicus, tragicus obliquus auriculae** and **transversus auriculae** are usually present in hylobatids, as is normally the case in modern humans. He suggested that the **incisurae terminalis** ('incisurae Santorini') and the '**intercartilagineus**' are usually absent in hylobatids, but that the **pyramidalis auriculae** ('trago-helicinus') is usually present in these primates. He stated that the **depressor helicis** is present in *H. moloch*, but is inconstant in *H. syndactylus* and missing in *H. lar*; we found this muscle in *H. muelleri* (Fig. 4), and Ruge (1911), Loth (1931) and Edgeworth (1935) also illustrated the muscle in *H. leuciscus* and *H. syndactylus*. Apart from the 'trago-helicinus' (pyramidalis auriculae *sensu* the present study)

Ruge (1911) also described (e.g., his Fig. 2) a muscle 'helicinus' in *H. leuciscus* and *H. syndactylus*, which seems to correspond to part of the helicis *sensu* the present study. Ruge (1911) also reported muscles 'auriculares proprii' (see e.g., his Fig. 4; which seem to include the transversus auriculae *sensu* the present study), as well as a muscle antitragicus in *H. leuciscus* and *H. syndactylus*. The **mandibulo-auricularis** is not present as a distinct, fleshy muscle in hylobatids; it possibly corresponds to part or the totality of the stylomandibular ligament, as is usually the case in modern humans. Seiler (1971d, 1976) suggested that a '**risorius**' might be occasionally present in hylobatids but as explained by Diogo et al. (2009b) at least some of the structures shown by this author in hylobatids clearly seem to be neither homologous to each other, nor to the risorius of *Homo*, *Pan* and *Gorilla*. For instance, Seiler (1971d: p. 362) describes a risorius in *H. agilis*, *H. lar* and *H. syndactylus*, and he also shows this muscle in his Fig. 591 of a *H. agilis* specimen and in Fig. 592 of a *H. syndactylus* specimen. However, it is not clear if the 'risorius muscles' of these taxa are homologous structures because the muscle of *H. syndactylus* runs much more horizontally than that of *H. agilis*, having a general configuration that is similar to that of the zygomaticus major (i.e., the muscle of *H. agilis* could be differentiated from the platysma myoides, while that of *H. syndactylus* could be differentiated from the zygomaticus major). In fact, to our knowledge all the other authors that have studied in detail the facial muscles of *Hylobates* did not find a distinct risorius muscle in any specimens of this genus; moreover, we also did not found this muscle in any of the hylobatid specimens dissected by us. Even if some of the structures described in these primates by Seiler (1971d, 1976) are homologous to the risorius of *Homo*, *Gorilla* and *Pan*, a risorius would nevertheless still represent an extremely rare condition within the Asian apes .

Zygomaticus major (Figs. 1, 3, 4, 6)

- Usual attachments: From the temporalis fascia and the zygomatic arch/bone (not from the ear, although it originates nearer to the ear than is usually the case in humans) to the corner of the mouth.
- Usual innervation: Branches of CN7.
- Notes: Within hylobatids, anteriorly the zygomaticus major often splits around the levator anguli oris facialis/depressor anguli oris, as reported by Deniker (1885), Hartmann (1886), Ruge (1911), Huber (1930b, 1931), and Edgeworth (1935), blending with these two muscles as well as with the orbicularis oris and the zygomaticus minor (e.g., Figs. 5, 6 and 7 of Ruge 1911). In the hylobatids dissected by us the zygomaticus major lies in the same plane as and is superior to the platysma cervicale and platysma myoides, as shown in the illustrations of Huber (1930b, 1931) i.e., it does not pass superficially to these latter muscles (as shown in the illustrations of Ruge 1911, Loth 1931 and Edgeworth 1935). Ruge (1911) describes and shows (e.g., his Fig. 2) a small 'auriculolabialis inferior'

(or 'platysma auricularis') in *H. moloch* and *H. syndactylus* that originates from the inferior margin of the ear and then blends with the platysma cervicale, lying near to, but is apparently not continuous with the zygomaticus major; it is not clear if this 'auriculolabialis inferior' is just a portion of the platysma cervicale or if this is instead the vestige of an attachment of the 'auriculolabialis inferior' (zygomaticus major *sensu* the present study) onto the ear, as found in various mammalian taxa.
  - Synonymy: Zygomaticus and possibly auriculolabialis inferior (Ruge 1911); part or totality of zygomatico-labialis (Edgeworth 1935); zygomaticus inferior (Seiler 1971d, 1976).

Zygomaticus minor (Figs. 1, 3, 4, 6)
  - Usual attachments: Mainly from the zygomatic bone and the orbicularis oculi (relatively far from the ear and near to the eye, as is usually the case in humans: see, e.g., Plate 26 of Netter 2006) to the corner of the mouth and to the upper lip, being mainly superficial (lateral) to the levator anguli oris facialis.
  - Usual innervation: Branches of CN7.
  - Notes: Ruge (1911) describes and shows (e.g., his Fig. 3) a small 'auriculolabialis superior' in *H. moloch* and *H. syndactylus*, which originates near the superior margin of the ear and then 'disappears', lying near to, but apparently not being continuous with, the posterior origin of the zygomaticus major and the zygomaticus minor; it is not clear if this is a vestige of an attachment of the 'auriculolabialis superior' (zygomaticus minor *sensu* the present study) onto the region of the ear as is the case in various mammalian taxa.
  - Synonymy: Temporolabialis and possibly auriculolabialis superior (Ruge 1911); zygomaticus superior (Seiler 1971d, 1976).

Frontalis (Fig. 4)
  - Usual attachments: From the galea aponeurotica to the skin of the eyebrow and the nose.
  - Usual innervation: Branches of CN7.

Auriculo-orbitalis (Figs. 1, 4)
  - Usual attachments: From the anterior portion of ear to the region of the frontalis.
  - Usual innervation: Branches of CN7.
  - Notes: In Terminologia Anatomica (1998) the **temporoparietalis** is considered to be a muscle that is usually present in modern humans (originating mainly from the lateral part of the galea aponeurotica and passing inferiorly to insert onto the cartilage of the auricle, in an aponeurosis shared with the other auricular muscles). However, according to authors such as Loth (1931) the temporoparietalis is usually absent as a distinct muscle in modern humans. According to Diogo et al. (2008, 2009b), the temporoparietalis and **auricularis anterior** derive from the auriculo-orbitalis so when the temporoparietalis is not

present as a distinct muscle these authors use the name auriculo-orbitalis to designate the structure that is often designated in the literature as 'auricularis anterior': that is, one can only use this latter name when the temporoparietalis is present. In various primates reported by Seiler (1976) including hominoids such as *H. moloch* and *Pongo pygmaeus*, he reported both a 'pars orbito-temporalis of the frontalis' and an 'auricularis anterior' attaching posteriorly onto the ear (N.B., he stated that the 'auricularis anterior is missing in *H. lar* and *H. syndactylus*). That is, the structure that he designated as 'auricularis anterior' is differentiated from the auriculo-orbitalis, as is the case in various other primates, but according to him contrary to those primates (including *Pan troglodytes* and *Gorilla gorilla*: see, e.g., Fig. 143 of Seiler 1976) in hominoids such as *H. moloch* and *P. pygmaeus* there is 'still' a connection between the main body of the auriculo-orbitalis and the ear. One would think that the temporoparietalis of taxa such as modern humans would probably correspond to those remaining fibers of the auriculo-orbitalis that did not differentiate into the 'auricularis anterior'; this was what Jouffroy and Saban (1971) suggested in their study on the facial muscles of mammals. However, in at least some, if not all, taxa this is clearly not the case, as shown for instance in Fig. 143 of Seiler (1976): the structure that he designates as 'pot', which corresponds to the remaining fibers of the auriculo-orbitalis that do not form an 'auricularis anterior' does not correspond to the structure that is usually designated as temporoparietalis in modern human anatomical atlases, which usually runs mainly superoinferiorly from the parietal bone to the temporal region, as its name indicates. Therefore, in order to be as consistent as possible with our previous studies and because the 'pars orbito-temporalis of the frontalis' and the 'auricularis anterior' *sensu* Seiler 1976 are very likely derived from the same anlage and are often related to each other, being often even continuous, we simply consider these two structures as parts/bundles of the auriculo-orbitalis *sensu* the present study (see, e.g., Fig. 74 of Seiler 1976). That is, the 'auricularis anterior' *sensu* Seiler 1976 is considered to be part of the auriculo-labialis *sensu* the present study, except in those few primates that have a distinct temporoparietalis (in those few primates the 'auricularis anterior' *sensu* Seiler 1976 is named by us as auricularis anterior, to clearly indicate that those primates have a distinct temporoparietalis). It is however possible, and in our opinion likely, that as suggested by Jouffroy and Saban (1971) the temporoparietalis of modern humans corresponds to the 'pars orbito-temporalis of the frontalis' *sensu* Seiler (1976) and that most of the confusion related to this issue is due to erroneous descriptions of the modern human temporoparietalis in anatomical atlases (which suggest that this is mainly a vertical muscle running superoinferiorly from the parietal bone to the temporal region, a description that does not match with the usual configuration of Seiler's 'pars orbito-temporalis of the frontalis' in other primates, which mainly runs horizontally to the region of the orbit to the region of the ear). If this is the case, then most primates would have a distinct temporoparietalis and a distinct auricularis anterior because both a 'pars orbito-temporalis of the frontalis' and an 'auricularis anterior' were reported in most primates by Seiler 1976. If this is so, then the 'pars orbito-temporalis of the frontalis' and the 'auricularis anterior' of all those primates should be designated as temporoparietalis and as auricularis anterior, respectively—we plan to address this subject in detail in future studies.

- Synonymy: Part of auriculaire antéro-supérieur (Deniker 1885); part or totality of the auricularis antero-superior (Ruge 1911); orbito-auricularis (Huber 1930b, 1931); auricularis anterior and probably part or totality of pars orbito-temporalis of frontalis (Seiler 1976).

Auricularis superior (Figs. 1,4)

- Usual attachments: From the superior margin of the ear to the galea aponeurotica.
- Usual innervation: Branches of CN7.
- Notes: Deniker (1885) states that in the fetal *Hylobates* sp. specimen dissected by him there is a single muscle ('auriculaire antéro-supérieur') instead of the three muscles (auricularis anterior, auricularis superior and temporoparietalis) found in the fetal gorilla that he dissected; this muscle seems to correspond to the auricularis superior plus the auriculo-orbitalis *sensu* the present study (he states that the temporoparietalis is missing in the fetal gibbon). Ruge (1911) describes and illustrates an 'auricularis superior primitivus' and an 'auricularis antero-superior' in *H. moloch* and *H. syndactylus*; these structures are partially blended with each other and correspond, respectively, to the auricularis superior and auriculo-orbitalis *sensu* the present study.
- Synonymy: Auricularis superior primitivus (Ruge 1911).

Orbicularis oculi (Figs. 1, 2, 4)

- Usual attachments: From a continuous bony attachment around the orbit to skin near the eye. It is usually divided into a pars palpebralis and a pars orbitalis, as in modern humans.
- Usual innervation: Branches of CN7.
- Synonymy: Orbiculaire des paupières (Deniker 1885); orbicularis palpebrarum (Hartmann 1886); orbicularis oculi superior et inferior (Seiler 1976).

Depressor supercilii (Figs. 1, 2, 3, 4)

- Usual attachments: From the ligamentum palpebrale mediale to the eyebrow.
- Usual innervation: Branches of CN7.
- Notes: Deniker (1885) does not describe a depressor supercilii in the fetal *Hylobates* sp. specimen dissected by him, and the muscle is also not shown in the adult *H. pileatus* illustrated in the figs. of Huber (1930b, 1931). However, the muscle was found by authors such as Ruge (1911) and Seiler (1976), and by

us in specimens of various hylobatid species, so it is probably usually present in most, or all, hylobatid species.

Corrugator supercilii (Fig. 4)

- Usual attachments: From the medial end of the supercilliary arch and from the dorsomedial 'roof' of the orbit to the eyebrow region.
- Usual innervation: Branches of CN7.
- Notes: The corrugator supercilii is not shown in the adult *H. pileatus* illustrated in the figs. of Huber (1930b, 1931). However, the muscle was found by authors such as Deniker (1885), Ruge (1911) and Seiler (1976), and by us, in specimens of various species of *Hylobates*, so it is probably usually present in most or all hylobatid species.
- Synonymy: Sourcilier (Deniker 1885).

Levator labii superioris (Figs. 1, 2, 3, 4)

- Usual attachments: From the infraorbital region mainly to the nose, although a few fibers may also attach onto the superior portion of the medial region of the upper lip.
- Usual innervation: Branches of CN7.
- Notes: Deniker (1885) states that in the fetal *Hylobates* sp. specimen dissected by him the levator labii superioris ('muscle profond') originates from the region that lies inferiorly to the suborbital foramen and inserts onto the ala of the nose being much less vertically-orientated than in modern humans (see Fig. 1 of his Plate 27); the levator labii superioris alaeque nasi ('muscle superficiel') runs from the orbital region to the ala of the nose and to the region of the upper lip. Ruge (1911: *H. moloch, H. syndactylus*) and Huber (1930b, 1931: *H. pileatus*) describe and illustrate a 'maxillo-naso-labialis' (levator labii superioris *sensu* the present study) 'primitively' running mainly from the nose to the region lying inferiorly to the inferolateral portion of the orbicularis oculi (e.g., Fig. 5 of Ruge 1911; Fig. 33 of Huber 1930b; Fig. 8 of Huber 1931). The descriptions of Seiler (e.g., Table 2 of Seiler 1970; Fig. 5224 of Seiler 1971d) and our own dissections (e.g., Figs. 1, 2) also support the idea that in hylobatids the levator labii superioris is not as markedly superoinferiorly directed as that of non-hylobatid catarrhines; it mainly runs, instead, posteroanteriorly and lateromedially from the infraorbital region to the nose, the muscle is thus strikingly similar to that found in non-catarrhine primates.
- Synonymy: Pars profunda des releveurs communs de l'aie du nez et de la lèvre superieure (Deniker 1885); maxillo-naso-labialis (Ruge 1911, Huber 1930b, 1931).

Levator labii superioris alaeque nasi (Figs. 1, 2, 3)

- Usual attachments: From the region of the ligamentum infraorbitale mediale to the upper lip and sometimes to the ala of the nose.
- Usual innervation: Branches of CN7.

- Notes: In the *H. lar* and *H. gabriellae* specimens dissected by us the levator labii superioris alaeque nasi is exactly as shown by authors such as Huber (1930b, 1931) and thus corresponds to the levator labii superioris alaeque nasi and likely also to at least part of the **'depressor glabellae'** shown in Fig. 251 of Seiler (1971c). Inferiorly the levator labii superioris attaches onto the upper lip (not on the nose) being mainly blended with the depressor septi nasi, the orbicularis oris and the levator labii superioris; superiorly the muscle seems to have a bony attachment to the region lying medially to the eye, blending mainly with the procerus, the frontalis, the depressor supercilii and the orbicularis oculi. See also the levator labii superioris above.
- Synonymy: Pars superficialis des releveurs communs de l'aie du nez et de la lèvre superieure (Deniker 1885); levator alae nasi (Hartmann 1886); levator labii superioris et nasi (Ruge 1911); part or totality of naso-labialis (Edgeworth 1935); part of caput angulare of quadratus labii superioris (Jouffroy and Saban 1971); levator labii superioris alaeque nasi and possibly also part or the totality of depressor glabellae (Seiler 1971c, 1976).

Procerus (Figs. 3, 4)
- Usual attachments: From the frontalis to the medial region of the nose.
- Usual innervation: Branches of CN7.
- Notes: According to Seiler (1971c, 1976) the procerus and the 'depressor glabellae' are usually present as distinct structures in hylobatids. However, the 'depressor glabellae' is often considered in the literature as part of the procerus (see e.g., Terminologia Anatomica 1998) and at least in the case of hylobatids it may well also correspond to part of the levator labii superioris alaeque nasi (see above).
- Synonymy: Pyramidal du nez (Deniker 1885); levator alae nasi (Hartmann 1886); procerus nasi or depressor glabellae (Huber 1930b, 1931); part or totality of naso-labialis (Edgeworth 1935); part of caput angulare of quadratus labii superioris (Jouffroy and Saban 1971); procerus and possibly part or totality of depressor glabellae (Seiler 1971c, 1976).

Buccinatorius (Figs. 2, 3)
- Usual attachments: From the pterygomandibular raphe, the infero-lateral surface of the maxilla, the fossa buccinatoria and the alveolar border of the mandible, mainly to the angle of the mouth and the upper and lower lips.
- Usual innervation: Branches of CN7.
- Synonymy: Part of labio-buccalis (Ruge 1911).

Nasalis (Figs. 2, 4)
- Usual attachments: From the maxilla, deep to the orbicularis oris, to the inferior portion of the lateral margin and the lateral portion of the inferior margin of the ala of the nose.
- Usual innervation: Branches of CN7.

- Notes: Seiler (1970, 1971c, 1976) describes a 'nasalis' and a **'subnasalis'** in various catarrhines, including hylobatids. The 'subnasalis' and 'nasalis' *sensu* Seiler could correspond to the pars alaris and pars transversa of the nasalis of modern human anatomy, respectively (compare e.g., Fig. 141 of Seiler 1976 to Plate 26 of Netter 2006). However, according to Seiler the 'subnasalis' is usually missing in *Homo* and *Gorilla*, so this seems to indicate that it does not correspond to the pars alaris of the nasalis *sensu* the present study, because this latter structure is usually found in modern humans. Be that as it may, as other authors do not refer to a 'subnasalis' muscle and as we also did not found a distinct separate 'subnasalis' in the hylobatids dissected by us, this structure probably corresponds to part of the nasalis and/or of the orbicularis oris *sensu* the present study. Seiler (1970, 1971c, 1976), also describes a 'depressor septi nasi' in various catarrhines, including hylobatids; he describes and shows a **'musculus nasalis impar'** in a few catarrhines, including *H. agilis* and *H. syndactylus* (according to him this structure is inconstant in these two species and missing in *H. lar, H. moloch* and *H. leucogenys*—thus according to Seiler the muscle is usually missing in hylobatids). One could hypothesize that this 'nasalis impar' could correspond to part or the totality of the depressor septi nasi that is illustrated in a few atlases of modern human anatomy as a vertical muscle that lies on the midline and that attaches mainly onto the inferomesial margin of the nose (compare, e.g., Fig. 145 of Seiler 1976 with Plate 26 of Netter 2006). However, most atlases of modern human anatomy show two depressor septi nasi muscles, one in each side of the body, running obliquely (superomedially) from the upper lip to a more medial part of the inferior region of the nose; that is, the 'nasalis impar' *sensu* Seiler (1975) does seem to correspond to an additional midline muscle that is only inconstantly present in catarrhines, while the 'depressor septi nasi' *sensu* Seiler (1976) is effectively similar to the depressor septi nasi shown in most atlases of modern human anatomy. Table 3 of Seiler (1979b) refers to a **'labialis superior profundus'** showing that the 'depressor septi nasi' and the 'labialis superior profundus' are not the same structure according to him, while Lightoller (1928a, 1934) does suggest that these are the same structure (Seiler 1976 clearly shows both these structures in various primates, e.g., in his Fig. 145 of *Gorilla*, the 'labialis superior profundus' probably corresponding to part of the orbicularis oris *sensu* the present study). It should be noted that Jouffroy and Saban (1971) call the nasalis as the 'naso-labialis profundus pars anterior' and the depressor septi nasi as the 'naso-labialis profundus pars mediana'; this seems to indicate that these two muscles derive from the same structure; in their Fig. 471, they designate the 'labii profundus superioris' as depressor septi nasi, thus suggesting that the 'labii profundus superior' could correspond to the depressor septi nasi of modern humans as suggested by Lightoller (1928a, 1934). However, our comparisons and dissections point out that the homologies proposed by

Seiler (1976) are somewhat doubtful and that: 1) the 'nasalis' *sensu* Seiler likely corresponds to part or the totality of the nasalis *sensu* the present study; 2) the pars transversa of the nasalis of modern humans might correspond to the 'subnasalis' *sensu* Seiler (see, e.g., Fig. 141 of Seiler 1976; however, Seiler stated that the 'subnasalis' is missing in modern humans), to the depressor septi nasi *sensu* Seiler 1976 (see, e.g., Fig. 145 of Seiler 1976), and/or to part of the nasalis *sensu* the present study (see, e.g., Fig. 145 of Seiler 1976); 3) the depressor septi nasi of modern humans does seem to correspond to the 'depressor septi' nasi *sensu* Seiler; therefore, the 'depressor septi nasi' *sensu* Seiler does not seem to correspond to part of the 'labialis superior profundus' *sensu* Seiler, as suggested by Lightoller (1928a, 1934; see, e.g., Fig. 141 of Seiler 1976), nor to the 'nasalis impar' *sensu* Seiler (in fact, Seiler stated that the 'nasalis impar' is missing in modern humans, which do usually have a depressor septi nasi *sensu* the present study). Deniker (1885) did not find a pars alaris nor a pars transversa of the nasalis in the fetal *Hylobates* sp. specimen dissected by him and it is not clear if these two bundles of the nasalis are, or not, usually differentiated in adult hylobatids.
- Synonymy: Part of labio-buccalis (Ruge 1911).

Depressor septi nasi (Fig. 2)
- Usual attachments: From the maxilla, deep to the orbicularis oris, to the inferior region of the nose.
- Usual innervation: Branches of CN7.
- Notes: Deniker (1885) does not describe a depressor septi nasi in the fetal *Hylobates* sp. specimen dissected by him, but in our *H. lar* specimen HU HL1 there is a depressor septi nasi *sensu* Fig. 4 of Seiler (1970), which is mainly superficial (i.e., anterior) to the nasalis; laterally it blends with the orbicularis oris, but medially it is oriented superomedially to attach onto the inferomesial region of the nose (Fig. 2). The depressor septi nasi of HU HL1 thus conforms with the hylobatid type of depressor septi nasi indicated in table 1 of Seiler 1970 (but see nasalis above).
- Synonymy: Part of labio-buccalis (Ruge 1911).

Levator anguli oris facialis (Fig. 2)
- Usual attachments: From the canine fossa of the maxilla to the angle of mouth.
- Usual innervation: Branches of CN7.
- Notes: As proposed by Diogo et al. (2008) and Diogo and Abdala (2010), we use the name levator anguli oris facialis here (and not the name 'levator anguli oris', as is usual in atlases of modern human anatomy) to distinguish this muscle from the **levator anguli oris mandibularis** (which is usually also designated as 'levator anguli oris' in the literature) found in certain reptiles, which is part of the mandibular (innervated by CN5) and not of the hyoid

(innervated by CN7) musculature. In the *H. lar* and *H. gabriellae* specimens dissected by us the levator anguli oris facialis is similar to that shown in Fig. 33 of Huber (1930b); inferolaterally it attaches onto the corner of the mouth, being blended with the orbicularis oris and particularly with the depressor anguli oris and is partially superficial (or lateral) to the zygomaticus major. In fact, the depressor anguli oris and the levator anguli oris facialis are practically continuous, thus supporting the idea that these two muscles are associated phylogenetically and ontogenetically; then the levator anguli oris facialis extends superomedially passing mainly deep to the zygomaticus minor, having a bony attachment to the canine fossa. Some of the deepest fibers of the muscle blend with the buccinatorius; this also supports the hypothesis that these two muscles are associated ontogenetically and phylogenetically.

- Synonymy: Caninus (Deniker 1885, Ruge 1911, Huber 1930b, 1931); part of caninus (Seiler 1976).

Orbicularis oris (Figs. 1, 2, 3)

- Usual attachments: From skin, fascia and adjacent regions of lips to skin and fascia of lips.
- Usual innervation: Branches of CN7.
- Notes: Seiler (1970, 1971c) describes a '**cuspidator oris**' in hylobatids. As suggested by him, this structure, which was designated as '**labialis superior profundus**' by Seiler 1976, probably corresponds to the '**incisivus labii superior**' *sensu* Lightoller (1928, 1934, 1939) and, thus to part of the orbicularis oris *sensu* the present study. Seiler (1976) also describes a '**labialis inferior profundus**' in hylobatids, which thus probably corresponds to the '**incisivus labii inferioris**' *sensu* Lightoller (1928, 1934, 1939) and to part of the orbicularis oris *sensu* the present study. Deniker (1885) states that in the fetal *Hylobates* sp. specimen dissected by him there are no distinct 'incisive' muscles. Ruge (1911) describes, and shows, a 'labio-buccalis' in *H. leuciscus* and *H. syndactylus*, which includes the orbicularis oris and the buccinatorius and apparently also includes part or the totality of the nasalis and of the depressor septi nasi *sensu* the present study (see, e.g., his Figs. 8 and 9).
- Synonymy: Part of labio-buccalis (Ruge 1911).

Depressor labii inferioris (Figs. 1, 3)

- Usual attachments: From the platysma myoides and the mandible to the lower lip.
- Usual innervation: Branches of CN7.
- Notes: Some authors (e.g., Ruge 1911 and Huber 1930b, 1931) have suggested that in hylobatids the depressor labii inferioris is not present as a distinct muscle or it is not as differentiated in these primates as it is in non-hylobatid hominoids such as modern humans. However, the descriptions of Deniker (1885) and Seiler (1976) and our own dissections show that the depressor labii

inferioris is often present as a distinct muscle in hylobatids. The depressor labii inferioris usually arises from the mandible and inserts onto the skin of the lower lip, being mainly blended with its counterpart as well as with the orbicularis oris, depressor anguli oris, mentalis and platysma myoides (it is mainly distinct from the platysma myoides because, in lateral view, its fibers run more vertically than the anterior fibers of this latter muscle).

- Synonymy: Carré des lèvres (Deniker 1885).

Depressor anguli oris (Figs. 1, 3)
- Usual attachments: From the angle of mouth to the fascia of the platysma myoides.
- Usual innervation: Branches of CN7.
- Notes: In hylobatids the depressor anguli oris usually does not reach the inferior margin of the mandible and does not extend inferomedially to meet its counterpart in the ventral midline, i.e., it does not form a **transversus menti** (e.g., Deniker 1885; our observations).
- Synonymy: Triangularis (Deniker 1885, Huber 1931, Edgeworth 1935); part of caninus (Seiler 1976).

Mentalis (Figs. 2, 3)
- Usual attachments: From the mandible to the skin below the upper lip, being mainly deep to the orbicularis oris and to the depressor labii inferioris.
- Usual innervation: Branches of CN7.
- Notes: In our H. *lar* specimen HU HL1 some of the fibers of the mentalis seem to contact those of its counterpart.
- Synonymy: Houppe du menton (Deniker 1885).

3.3 Branchial musculature

Stylopharyngeus (Fig. 13)
- Usual attachments: This muscle has almost never been described in detail in hylobatids; it was reported in *H. hoolock* by Kanagasuntheram (1952–1954) and in *H. syndactylus*, *H. agilis* and *H. moloch* by Kohlbrügge (1890–1892). In our *H. gabriellae* specimen VU HG1 the stylopharyngeus runs from the tympanic region (there is no well-developed, ossified styloid process) to the pharynx, passing between the middle and superior constrictors.
- Usual innervation: Glossopharyngeal nerve (Kohlbrügge 1890–1892: *H. syndactylus*, *H. agilis*, *H. moloch*).
- Notes: The **ceratohyoideus** and **petropharyngeus** are not present as distinct muscles in hylobatids.

Trapezius (29.0 g; Figs. 10, 15, 21, 22, 23)
- Usual attachments: From the vertebral column and the thoracolumbar fascia to the spine and to the acromion of the scapula and to more than the lateral 1/3 of the clavicle.

- Usual innervation: According to Kohlbrügge (1890–1892) innervation is usually by XI and C4 in *H. agilis* and *H. syndactylus*, but in *H. moloch* it seems to be by C2, C3 and XI; according to Schück (1913b) in *H. syndactylus* it is usually by XI and C3 and C4 (in three specimens of this species) but sometimes is by XI and C2 and C4 (in one specimen of the same species).
- Notes: According to Gibbs' (1999) literature review in hylobatids the trapezius has no occipital origin (such an origin was also lacking in the fetal *Hylobates* sp. specimen dissected by Deniker 1885), the cervical origin begins at C5, and there is a nuchal ligament (as described by, e.g., Plattner 1923 and Sonntag 1924). However, Donisch (1973) reported a female *H. lar* specimen in which this latter ligament was "poorly developed" and Andrews and Groves (1976) stated that in hylobatids the most cranial origin is usually from C2–C3, and that in *H. syndactylus* it is from the occipital region, the most caudal origin being usually from T7 or T10 (according to them in *H. syndactylus* the most cranial and caudal origins of the trapezius are from the occipital region and T9–10 respectively, in *H. hoolock* from C3 and T3–7, in *H. muelleri* from C2 and thoracic vertebrae, in *H. moloch* from C2 and T2–7, in *H. agilis* from C2 and T2–7, and in *H. lar* the cranial origin is from cervical ribs). Kohlbrügge (1890–1892) stated that in *H. moloch* and *H. agilis* the trapezius has an upper portion originating from the ligamentum nuchae and C2–C6 and a lower portion originating from C7–T9, but that in *H. syndactylus* the muscle also originates from the occipital region and does extend down to T9 and T10, while in *H. agilis* it reaches T10; it inserts laterally to the levator claviculae, onto the lateral third of the clavicle, the acromion, and the scapular spine, but in *H. agilis* it inserted onto the middle of the clavicle instead. Schück (1913b) dissected four specimens of *H. syndactylus* and reported that the trapezius reached T12 in one specimen, T11 in two specimens, and T10 in one specimen. Sonntag (1924) stated that in *Hylobates* the trapezius has no occipital origin and arises from the lower cervical and upper seven thoracic spinous processes. Donisch (1973) reported a *H. lar* female in which the trapezius originated from the ligamentum nuchae and spinous processes of C7-T11 and inserted onto the lateral margin of the clavicle, the acromion and the scapular spine, and stated that Plattner (1923) described an occipital origin of the trapezius in *H. syndactylus*. Michilsens et al. (2009) dissected 11 hylobatids (3 *H. lar*, 2 *H. pileatus*, 2 *H. moloch* and 4 *H. syndactylus*) and stated that in all these specimens the trapezius connected the cervical and thoracic vertebrae (and not the occiput) to the lateral third of the clavicle, the scapular spine, and the acromion. In our *H. lar* specimen HU HL1 the trapezius does not attach onto the occiput, but does insert onto the acromion and spine of the scapula and onto the clavicle (onto the lateral 3.5 cm of the total 9 cm of this latter bone). In our *H. gabriellae* specimen VH HG1 the trapezius is not fused to the latissimus dorsi and does not reach the neurocranium, extending only to the ligamentum nuchae, at the level of the second or third cervical

vertebrae; it inserts onto the scapular spine, acromion and clavicle (onto the lateral 3 cm of the total 6 cm of this latter bone). In summary, although in a few hylobatids, particularly of the species *H. syndactylus*, there is a small origin of the trapezius from the cranium, as reported by Kohlbrügge (1890–1892), Plattner (1923), Miller (1932) and Andrews and Groves (1976), in the vast majority of these primates, including various specimens of *H. syndactylus*, there is no direct cranial origin (e.g., Deniker 1885, Kohlbrügge 1890–1892, Sonntag 1924, Loth 1931, Donisch 1973, Andrews and Groves 1976 and Michilsens et al. 2009; our dissections).

Sternocleidomastoideus (8.7 g; Figs. 5, 8, 9, 10)

- Usual attachments: The caput sternomastoideum runs from the sternum to the mastoid region and the lateral portion of the nuchal crest; the caput cleidomastoideum runs from the medial portion of the clavicle and, sometimes, also from a small portion of the sternum, to the mastoid region.
- Usual innervation: According to Kohlbrügge (1890–1892) it is by C2, C3 and C4 and XI in *H. moloch*, *H. agilis* and *H. syndactylus*, but of the 4 specimens of *H. syndactylus* dissected by Schück (1913b) it was innervated by XI and C2 and C3 in one specimen, by XI and C2 in two specimens, and by XI and C3 in one specimen.
- Notes: The **cleido-occipitalis** is usually not present as a distinct structure in hylobatids. According to Gibbs' (1999) literature review in hylobatids the origin of the sternocleidomastoideus extends down to the level of the notch for the third rib. Deniker (1885) described a fetal *Hylobates* sp. specimen with a sternocleidomastoideus divided into a caput sternomastoideum ('sterno-mastoïdo-occipital'; running from the sternum to the mastoid and occipital regions, its superior part being in turn divided into two bundles, corresponding to the cleido-occipitalis and to the sternomastoideus of other mammals) and a caput cleidomastoideum (running from the clavicle to the mastoid region, its insertion onto this region being deeper to that of the sternomastoideus); the caput cleidomastoideum was not perforated by the 'spinal nerve' (accessorius *sensu* the present study) contrary to the gorilla fetus dissected by him. Kohlbrügge (1890–1892) stated that in *H. moloch* and *H. agilis* the caput sternomastoideum runs from the manubrium to the temporal bone and that this head only reaches the occipital region in *H. syndactylus*; the caput cleidomastoideum runs from the sternal end of the clavicle to the region of the external acoustic meatus. Schück (1913ab) reported a *H. syndactylus* specimen in which the caput sternomastoideum is partially fused with the trapezius, through the extension of the cranial attachment of the caput sternomastoideum as far as the middle of the occiput; Fig. 7 of Schück (1913b) shows a specimen of *H. syndactylus* with 'accessory heads' of the sternocleidomastoideus, which do not seem to correspond to either the cleido-occipitalis or the cleidomastoideus

of other mammals. Stewart (1936) stated that the clavicular attachment of the cleidomastoideus is more distant from the sternal end of the clavicle in *Hylobates*, less in *Pan*, and still less in *Gorilla* (*Pongo* and *Homo* were not measured); the caput sternomastoideum originates from the manubrium, both fleshy and tendinously, together with its counterpart and inserts onto the mastoid process and the adjacent superior nuchal line through an aponeurosis, while the caput cleidomastoideum originates from the medial part of the clavicle (but does not reach its sternal end) and inserts onto the mastoid process deep to the anterior border of the caput sternomastoideum. In our *H. lar* specimen HU HL1 superiorly the caput cleidomastoideum and the caput sternomastoideum seem to be more separated than in modern humans; inferiorly the caput cleidomastoideum (attached onto the clavicle) is somewhat distant (1 or 2 cm) from the sternum (the caput sternomastoideum attaches onto the sternum, but it was not possible to check if it contacted its counterpart at the midline); superiorly the caput cleidomastoideum attaches onto the mastoid process and the caput sternomastoideum attaches onto the mastoid process and the adjacent superior nuchal line. In our *H. gabriellae* specimen VU HG1 the caput sternomastoideum is broader and more superficial, running from the sternum and the mesial 1 to 1.5 cm of the clavicle to the mastoid and occipital regions; the caput cleidomastoideum is deeply blended with this former head, running from the mesial 2 cm of the total 6 cm of the clavicle, to the mastoid region.

Constrictor pharyngis medius (Fig. 13)

- Usual attachments: This muscle has almost never been described in detail in hylobatids. Kanagasuntheram (1952–1954) illustrated the middle constrictor in *H. hoolock* and Kohlbrügge (1890–1892) reported *H. syndactylus*, *H. agilis*, and *H. moloch* specimens in which this muscle inserts onto the body and greater horn of the hyoid bone (so there is a pars ceratopharyngea, but no pars chondropharyngea, unless one considers that the portion of the muscle that inserts onto the body of the hyoid bone includes the pars chondropharyngea). Clegg (2001) stated that in the *H. muelleri* specimen dissected by her the middle constrictor attaches only to the greater horn of the hyoid bone. In our *H. gabriellae* specimen VU HG1 the muscle seemingly inserts exclusively onto the greater horn of the hyoid bone (pars ceratopharyngea); there is seemingly no distinct pars chondropharyngea.
- Usual innervation: Data are not available.

Constrictor pharyngis inferior (Fig. 14)

- Usual attachments: This muscle was almost never described in detail in hylobatids. Kanagasuntheram (1952–1954) illustrated the inferior constrictor in *H. hoolock*, and Kohlbrügge (1890–1892) reported *H. syndactylus*, *H. agilis* and *H. moloch* specimens in which the muscle inserted onto the thyroid (pars thyropharyngea) and cricoid (pars cricopharyngea) cartilages, as found in our *H. gabriellae* specimen VU HG1. Saban (1968) stated that in *Hylobates* the muscle does not have a dorsal median raphe.

- Usual innervation: Both the superior laryngeal and recurrent laryngeal nerves (Kohlbrügge 1890–1892: *H. moloch, H. agilis, H. syndactylus*).

Cricothyroideus (Fig. 14)

- Usual attachments: From the cricoid cartilage to the thyroid cartilage, specifically to its inferior and inferomesial portion as well as to a small portion of the lateral margin of its inferior horn.
- Usual innervation: According to Kohlbrügge (1890–1892: *H. moloch, H. agilis, H. syndactylus*) it is by the external branch of the superior laryngeal nerve; according to Kanagasuntheram (1952–1954) and Saban (1968) in *H. hoolock* the 'pars externa' (see notes below) of the cricothyroideus is innervated by the external branch of the superior laryngeal nerve, while the pars interna is innervated by the recurrent nerve.
- Notes: Starck and Schneider (1960) suggested that they did not found a distinct pars obliqua and a distinct pars recta in their specimens of *H. syndactylus*, but these two portions were reported in other specimens of this and other hylobatid species (e.g., Kohlbrügge 1890–1892, Kanagasutheram 1952–1954 and Saban 1968) and were also found in the hylobatids dissected by us (Fig. 14). A pars interna of the cricothyroideus is also present in hylobatids according to Duckworth (1912), Kanagasuntheram (1952–1954), Saban (1968), Kohlbrügge (1890–1892, 1896) and to our dissections, but not according to Starck and Schneider (1960). In fact, Kohlbrügge (1890–1892) reported *H. syndactylus*, *H. agilis* and *H. moloch* specimens in which the cricothyroideus runs from the cricoid cartilage to the thyroid cartilage and is clearly divided into a pars interna, a pars recta, and a pars obliqua; the pars recta is more superficial (lateral) and more ventral than the pars obliqua and only a few fibers of the pars interna attach onto the lower (posterior) margin of the thyroid cartilage, most of the remaining fibers of this bundle extending anteriorly to attach onto the inner side of the thyroid cartilage. Kohlbrügge (1890–1892) also suggests that the cricothyroideus contacts its counterpart at the ventral midline. Fig. 56 of Starck and Schneider (1960) shows a *H. syndactylus* specimen in which the cricothyroideus meets its counterpart at the ventral midline and seems to attach onto the lateral side of the inferior horn of the thyroid cartilage. Saban (1968) stated that, as described by Kanagasuntheram 1952–1954, in *Hylobates* the cricothyroideus has a peculiar configuration because it is divided into 4 bundles (i.e., the pars interna and a 'pars externa' that is divided into a deep subdivision and a superficial subdivision that is in turn subdivided into a 'pars interna' and a 'pars recta'). The 'pars externa' has 1) a superficial subdivision made up of anterior (ventral) fibers that are more transversely oriented (pars recta *sensu* the present study) and posterior (dorsal) fibers that are more oblique

(probably correspond to part of the pars obliqua *sensu* the present study) and that insert onto the thyroid cartilage and onto the posterior thyroid cornu and 2) a deeper subdivision made up of oblique fibers inserting onto the posterior part of the thyroid cartilage (probably correspond to part of the pars obliqua *sensu* the present study). The pars interna inserts onto the posterior margin of the thyroid cartilage; posteriorly (dorsally) the cricothyroideus has some transverse fibers that meet at the midline, forming the muscle thyroideus transversus (see below). In our *H. gabriellae* specimen VU HG1 the cricothyroideus is as shown in Plate XI of Kohlbrügge (1890–1892), i.e., there is a pars recta (more longitudinal and ventral), a pars obliqua (more oblique and dorsal) and also a pars interna going to the inner side of the thyroid cartilage; the cricothyroideus meets its counterpart at the ventral midline, and attaches onto a significant portion of the lateral side of the inferior horn of the thyroid cartilage.
- Synonymy: Cricothyreoideus anticus (Kohlbrügge 1890–1892, 1896).

Thyroideus transversus (Fig. 14)
- Usual attachments: See notes below.
- Usual innervation: External branch of the superior laryngeal nerve (Kohlbrügge 1890–1892: *H. moloch, H. agilis, H. syndactylus*).
- Notes: In hylobatids there is often a distinct muscle thyroideus transversus (also designated in the literature as 'thyroideus impar') that lies on the ventral margin of the larynx and runs transversely to connect the posteroventromedial portion of the two sides (e.g., Plate IX of Kohlbrügge 1890–1892, and Fig. 56A of Starck and Schneider 1960, and also Loth 1931, and Saban 1968; it should be noted that this muscle might be present as a rare anomaly in modern humans: e.g., Loth 1931). In our *H. gabriellae* specimen VU HG1 the thyroideus transversus is as shown in Plate XI of Kohlbrügge (1890–1892), being continuous with its counterpart at the ventral midline, originating from the cricoid cartilage and seemingly also from the first ring of the trachea; it seems to derive ontogenetically and phylogenetically from the cricothyroideus and thus from the anlage that also gives rise to the inferior constrictor.
- Synonymy: Thyroideus impar (Starck and Schneider 1960, Saban 1968).

Constrictor pharyngis superior
- Usual attachments: This muscle has almost never been described in detail in hylobatids. Kohlbrügge (1890–1892) reported *H. syndactylus, H. agilis,* and *H. moloch* specimens in which the superior constrictor is mainly made up of a pars glossopharyngea running from the tongue to the pharynx (i.e., the pars mylopharyngea); the pars buccopharyngea and pterygopharyngea are missing (he describes a bundle of the longitudinal layer of the muscles of the pharynx that is mainly originated from the pterygoid hamulus, which thus seems to correspond to the muscle pterygopharyngeus *sensu* the present study, thus supporting the idea that the fibers of this latter muscle may be

included in the superior constrictor of humans: see below). The illustrations of Kanagasuntheram (1952–1954) and Saban (1968) show a *H. hoolock* specimen in which there is a pterygopharyngeus separated from the superior constrictor and in which the superior constrictor seems to correspond to both the pars mylopharyngea and the pars glossopharyngea of modern humans. We could not discern if in our *H. gabriellae* specimen VU HG1 the superior constrictor has a pars buccopharyngea and/or a pars mylopharyngea (there is no pars pterygopharyngea because the pterygopharyngeus seems to be present as a distinct muscle: see below), but even if they were present they would be very thin because the pars glossopharyngea is well developed.

- Usual innervation: Data are not available.
- Synonymy: Mylo-glosso-pharyngeus (Saban 1968).

Pterygopharyngeus (Fig. 13)
- Usual attachments: In hylobatids the pterygopharyngeus is usually present as a distinct muscle (e.g., Kohlbrügge 1890–1892, Kanagasuntheram 1952–1954, Saban 1968, our dissections; see Fig. 285 of Saban 1968), running mainly from the pterygoid hamulus to the pharyngeal wall.
- Usual innervation: Data are not available.
- Notes: Authors such as House (1953) suggest that the pterygopharyngeus (e.g., of rats and hylobatids) corresponds to the pars pterygopharyngea of the constrictor pharyngis superior of modern humans, and eventually also to part or the totality of the pars buccopharyngea and mylopharyngea, unless these two latter structures are *de novo* formations; however, authors such as Edgeworth (1935) consider that such an hypothesis is questionable, because the orientation of the fibers of the pterygopharyngeus (e.g., rats) is much more similar to that of the palatopharyngeus (i.e., it is more cranio-caudaly then dorso-ventrally oriented). Therefore, although most authors defend (as we do in the present study) the proposal that the pterygopharyngeus of rats and hylobatids corresponds to part of the constrictor pharyngis superior of modern humans, until more data are available one cannot exclude the hypothesis that the pterygopharyngeus might be simply missing or eventually be deeply blended with the palatopharyngeus in modern humans and most other primates and/or that it might correspond to part of the palatopharyngeus of monotremes.
- Synonymy: Part of longitudinal layer of pharyngeal muscles (Kohlbrügge 1890–1892).

Palatopharyngeus
- Usual attachments: This muscle has almost never been described in detail in hylobatids. It was illustrated in *H. hoolock* by Kanagasuntheram (1952–1954) and Saban (1968), and Kohlbrügge 1890–1892 reported *H. syndactylus*, *H. agilis* and *H. moloch* specimens in which the palatopharyngeus connects the palatine region to the pharynx. In our *H. gabriellae* specimen VU HG1 the muscle runs

from the palatine region (not from the pterygoid hamulus) to the pharyngeal wall, being similar to that of modern humans. It was not possible to discern if this specimen has a '**sphincter palatopharyngeus**' (**palatopharyngeal sphincter**, or **Passavant's ridge**), as is often the case in modern humans (see, e.g., pl. 65 of Netter 2006).

- Usual innervation: Data are not available.

Salpingopharyngeus
- Usual attachments: See notes below.
- Usual innervation: Data are not available.
- Notes: We could not discern if the salpingopharyngeus is present, or not, in the hylobatids dissected by us. To our knowledge the salpingopharyngeus has almost never been reported in hylobatids, but one should keep in mind that very few authors have actually provided detailed descriptions of the pharyngeal muscles of these primates. One of the few exceptions is Kohlbrügge (1890–1892) who stated that part of the longitudinal layer of the pharyngeal muscles originates from the Eustachian tube in *H. syndactylus*, thus indicating that at least in the specimens of this species there are some fibers that seemingly correspond to the salpingopharyngeus of modern humans.

Musculus uvulae
- Usual attachments: This muscle has almost never been described in detail in hylobatids. Edgeworth (1935) and Saban (1968) suggested that it usually present in hominoids, but they did not refer specifically to hylobatids. In our *H. gabriellae* VU HG1 specimen the musculus uvulae seems to be present and Kohlbrügge (1890–1892) stated that in *H. syndactylus*, *H. agilis*, and *H. moloch* the 'levator uvulae' (musculus uvulae *sensu* the present study) is well developed.
- Usual innervation: Data are not available.

Levator veli palatini
- Usual attachments: This muscle was almost never described in detail in hylobatids. It was illustrated in *H. hoolock* by Kanagasuntheram (1952–1954) and Saban (1968) and Kohlbrügge 1890–1892 reported *H. syndactylus*, *H. agilis* and *H. moloch* specimens in which the levator veli palatini runs from the ventral spine of the superior tympanic bone (which is connected to the fascia of the pterygoideus medialis), the Eustachian tube, and the medial surface of the medial pterygoid processes to the palatine aponeurosis, contacting its counterpart at the midline. In our *H. gabriellae* specimen VU HG1 the levator veli palatini is more horizontal than in modern humans, running from the region of the Eustachian tube and adjacent areas of the neurocranium to the soft palate; it is clearly medial to the tensor veli palatini and lateral to the palatopharyngeus.
- Usual innervation: Data are not available.

Thyroarytenoideus (Fig. 14)
- Usual attachments: See notes below.
- Usual innervation: Recurrent laryngeal nerve (Kohlbrügge 1890–1892: *H. moloch, H. agilis, H. syndactylus*).
- Notes: With respect to hominoids and other primates there has been controversy regarding the homologies of the thyroarytenoid bundles and the presence/ absence of a distinct **musculus vocalis**. Kohlbrügge (1896) dissected gorillas, chimpanzees and orangutans as well as taxa such as *Cebus, Semnopithecus, Hylobates* and *Macaca*, and stated that he could not find a distinct attachment of the thyroarytenoideus onto a true vocal cord (such as that found in modern humans) in any of these taxa, except perhaps in *Pongo*; within all the taxa mentioned above, he found an attachment onto the cricoid cartilage in *Hylobates* and *Colobus*. Giacomini (1897) examined the larynx of a gorilla and of an *Hylobates lar* specimen and based on these results and on his previous studies, he stated that only in modern humans is there a distinct, well-developed, musculus vocalis directly connected to the vocal cord, although in at least some *Hylobates* and seemingly in some *Pan*, there are a few fibers of the thyroarytenoideus that are somewhat isolated and situated near the vocal cord (see, e.g., Fig. 2 of his Plate II). Duckworth (1912) examined specimens from all the five extant hominoid genera, as well as of *Macaca, Cebus, Semnopithecus* and *Tarsius* and suggested that a well-developed, distinct musculus vocalis associated with the plica vocalis is also only consistently present in modern humans, but that great apes, and particularly chimpanzees, do show a configuration that is somewhat similar to that found in modern humans, in that they have a poorly developed/differentiated musculus vocalis (see, e.g., his Figs. 24 and 17). According to him, in modern humans the superior portion of the thyroarytenoideus usually also forms a distinct structure that is often associated with the region of the ventriculus (which corresponds to the **'musculus ventricularis laryngis'** *sensu* Kelemen 1948, 1969), although a somewhat similar configuration is seen in gorillas and in chimpanzees. Loth (1931) argued that in non-hominoid primates the vocal cords are mainly formed by a well-developed folds of mucous membrane, which are not really in contact with the musculus vocalis. He stated that in hominoids such as *Pan* and *Gorilla* the size of the folds is smaller, but the musculus vocalis is also not really connected to these folds, such a connection being however found in most modern humans. Edgeworth (1935) defended the notion that a musculus vocalis is found in some primates and suggested that when this structure is present the thyroarytenoideus becomes a **'thryroarytenoideus lateralis'** because its inferior/mesial part gives rise to the vocalis muscle. Starck and Schneider (1960) described a 'pars lateralis' as well as a 'pars medialis' that usually goes to the vocal fold/cord in *Pan, Pongo* and *Gorilla*, but not in *Hylobates*;

the latter corresponding to the musculus vocalis of modern humans. They did not find a pars aryepiglottica or a pars thyroepiglottica of the thyroarytenoideus in *Hylobates* and *Pongo*, but stated that other authors did report at least one of these structures in *Gorilla* and *Pan*. Saban (1968) clarified the nomenclature of the thyroarytenoideus and suggested that this muscle may be divided into the following structures: 1) a pars superior (often designated as '**thyroarytenoideus superior**', 'thyroarytenoideus lateralis' or '**ventricularis**'); 2) a pars inferior (often designated as '**thyroarytenoideus inferior**', '**thyroarytenoideus medialis**', or musculus vocalis); 3) a '**ceratoarytenoideus lateralis**'; 4) a 'pars intermedia' (but this name was only used by a few authors such as Starck and Schneider 1960 who stated that some primates might have a pars superior, a pars inferior, and a pars intermedia); 5) a pars thyroepiglottica; 6) a pars aryepiglottica; 7) a pars arymembranosa; and 8) a pars thyromembranosa. According to Saban (1968) the 'ceratoarytenoideus lateralis' is usually fused with (but not differentiated from, as suggested in some anatomical atlases) the cricoarytenoideus posterior, being only a distinct muscle in a few taxa and, within primates, in *Pan* (and still in this case this seems to constitute a variant/anomaly) where it is a small muscle running from the dorsal face of the inferior thyrohyoid horn to the arytenoid cartilage. Also according to Saban the pars superior and pars inferior are more superior and inferior, respectively, in apes and modern humans (in modern humans the more inferior or medial part i.e., the pars inferior is well developed and often designated as musculus vocalis) and more lateral/medial in primates such as *Macaca* and *Papio*; in *Pongo* the pars inferior is well-developed and lies anterior to the vocal cord but is not associated with it; Aiello and Dean (1990) stated that in non-human hominoids the pars aryepiglottica is often reduced in size or absent. Kohlbrügge (1890–1892) reported *H. syndactylus*, *H. agilis*, *H. lar* and *H. moloch* specimens in which the thyroarytenoideus originates from the inner side of the ventral portion of the thyroid cartilage; it has a main body attaching onto the processus muscularis and the processus vocalis of the arytenoid cartilage with a small, inferior (and seemingly lateral: see Fig. 3 of his Plate XI) bundle going to the cricoid cartilage. This bundle, which he designated as 'portio thyreocricoidea' is present in all specimens except in two of the 4 specimens of *H. syndactylus* dissected by him. According to him the thyroarytenoideus does not attach onto the vocal cords/membranes, nor onto the epiglottis and there is also no 'superior thyroarytenoideus' such as that often found in modern humans; the portions of the thyroarytenoideus that go to the arytenoid and cricoid cartilages do not seem to correspond to the pars superior and pars inferior *sensu* the present study. Saban (1968) stated that in hylobatids the thyroarytenoideus has two sections, one is anterior and is associated with the vocal cord and the ventricule, and the other is posterior and associated to both the cricoid and arytenoid cartilages. These sections could correspond to the pars superior and pars inferior *sensu* the present study, but this does not seem to be the case because the pars inferior *sensu* the present study seemingly corresponds to the 'muscle vocalis' of other authors, while in the description of Saban 1968 it is the anterior portion that goes to the vocal cord and ventricle. In our *H. gabriellae* specimen VU HG1 the thyroarytenoideus is as shown in Plate XI of Kohlbrügge (1890–1892), i.e., there is a broader bundle going from the thyroid cartilage to the arytenoid cartilage and a thinner bundle going from the thyroid cartilage to the cricoid cartilage; these two bundles seem to correspond to the pars superior and pars inferior of other taxa; the 'ceratoarytenoideus lateralis', 'pars intermedia', pars aryepiglottica, pars thyroepiglottica, pars arymembranosa and pars thyromembranosa are seemingly not present as distinct structures. However, it is possible that one of the two bundles of label 11 of Fig. 56E of Starck and Schneider (1960; *H. syndactylus*) might correspond to a 'ceratoarytenoideus lateralis'. In summary, the descriptions of the thyroarytenoideus of hylobatids provided in the literature are somewhat confusing and even contradictory: 1) in Fig. 56 of Starck and Schneider (1960) the thyroarytenoideus seems to be mainly undivided; 2) Kohlbrügge (1890–1892) shows anterior and posterior portions going respectively to the arytenoid and cricoid cartilages and he stated that there is no 'thyroarytenoideus superior'; 3) Saban (1968) describes a posterior portion going to the cricoid and arytenoid cartilages and an anterior portion going to the vocal cord and ventricle; 4) Harrison (1995) states that at least some *Hylobates* have a distinct 'thyroarytenoideus superior'; 5) in the *H. gabriellae* specimen dissected by us there is an anterior portion going to the arytenoid cartilage and a posterior portion going to the cricoid cartilage, so this latter portion could correspond to the pars inferior of other taxa, but we cannot be completely sure about this.

Musculus vocalis (see thyroarytenoideus above)
- Usual attachments: See thyroarytenoideus above.
- Usual innervation: See thyroarytenoideus above.
- Notes: See thyroarytenoideus above.

Cricoarytenoideus lateralis
- Usual attachments: From the anterior portion of the cricoid cartilage to the arytenoid cartilage.
- Usual innervation: Recurrent laryngeal nerve (Kohlbrügge 1890–1892: *H. moloch*, *H. agilis*, *H. syndactylus*).

Arytenoideus (Figs. 12, 14)
- Usual attachments: From the ipsilateral arytenoid cartilage to the contralateral arytenoid cartilage being a continuous muscle without a clear median raphe; the **arytenoideus obliquus** and **arytenoideus transversus** are not present as distinct muscles (e.g., Kohlbrügge 1890–1892, Starck and Schneider 1960, Clegg 2001, our dissections).

- Usual innervation: Recurrent laryngeal nerve (Kohlbrügge 1890–1892: *H. moloch, H. agilis, H. syndactylus*).
- Synonymy: Interarytenoideus or arytenoideus transversus (Kohlbrügge 1890–1892, Starck and Schneider 1960).

Cricoarytenoideus posterior (Figs. 12, 14)

- Usual attachments: From dorsal portion of the cricoid cartilage to the arytenoid cartilage; there is usually no attachment onto the inferior horn of the thyroid cartilage i.e., there is no **ceratocricoideus** *sensu* Harrison (1995).
- Usual innervation: Recurrent laryngeal nerve (Kohlbrügge 1890–1892: *H. moloch, H. agilis, H. syndactylus*).
- Notes: Kohlbrügge (1890–1892) suggested that in the *H. muelleri, H. agilis* and *H. moloch* specimens dissected by him the cricoarytenoideus posterior and its counterpart covered the dorsal midline, but that in *H. syndactylus* and *H. lar* these structures were at least partially separated at the midline. However, it is not clear if this was a partial, or a complete, separation, because in the *H. syndactylus* specimen shown in Fig. 56 of Starck and Schneider (1960) the muscles of both sides do meet each other at the dorsal midline. Be that as it may, the contact between the muscles of the two sides does seem to be a common condition for hylobatids, as this occurs in the *H. muelleri, H. agilis* and *H. moloch* specimens dissected by Kohlbrügge (1890–1892) and in the *H. syndactylus* specimen reported by Starck and Schneider (1960) as well as in the single hylobatid specimen in which we could discern this feature in detail (*H. gabriellae* VU HG1).

3.4 Hypobranchial musculature

Geniohyoideus (0.6 g; Fig. 9)

- Usual attachments: From the mandible to the hyoid bone, lying close to its counterpart at the midline; its insertion onto the hyoid bone bifurcates the posterior portion of hyoglossus (i.e., that portion of this latter muscle that attaches onto the hyoid bone; see hyoglossus below).
- Usual innervation: CN12 (Kohlbrügge 1890–1892: *H. moloch, H. agilis, H. syndactylus*).
- Notes: Wall et al. (1994) stated that stimulation of the geniohyoideus in *H. lar* elicited slight mandibular depression and marked elevation of the hyoid with tongue protrusion.

Genioglossus (Figs. 9, 11)

- Usual attachments: This muscle has almost never been described in detail in hylobatids. It was illustrated in *H. hoolock* by Kanagasuntheram (1952–1954) and Saban (1968), in *H. lar* by Dubrul (1958) and in *H. syndactylus, H. agilis, H. moloch* and *H. lar* by Kohlbrügge (1890–1892). This latter author stated that it is a strong muscle originating from a recess above the mental spine of the mandible. He also described a *H. syndactylus* specimen with a 'muscle glosso-epiglotticus' reaching the epiglottis (see Fig. 4 of his Plate XI), but he stated that the 'glosso-epiglotticus' was not present in the *H. lar*, *H. agilis*, and *H. moloch* specimens dissected by him. In our *G. gabriellae* specimen VU HG1 the genioglossus is well separated from the geniohyoideus and also from its counterpart, at least its ventral portion, running from the mandible to the tongue and hyoid bone.

- Usual innervation: Data are not available.
- Notes: The muscles **genio-epiglotticus**, **glosso-epiglotticus**, **hyo-epiglotticus**, and **genio-hyo-epiglotticus**, described by authors such as Edgeworth (1935) and Saban (1968) in some primate and non-primate mammals, do not seem to be present as distinct structures in our *H. gabriellae* specimen VU HG1 (according to these authors these muscles are usually not present in catarrhine taxa).

Intrinsic muscles of tongue
- Usual attachments: To our knowledge, there are no detailed published descriptions of these muscles in hylobatids and we could not examine them in detail in our dissections. However, the **longitudinalis superior**, **longitudinalis inferior**, **transversus linguae** and **verticalis linguae** are consistently found in modern humans and at least some other primate and non-primate mammals, so these four muscles are very likely also present in hylobatids. Detailed studies of the tongue and its muscles in hylobatids are clearly needed.
- Usual innervation: Data are not available.

Hyoglossus (Figs. 11, 12)
- Usual attachments: In our *H. gabriellae* specimen VU HG1, the hyoglossus is differentiated into a **ceratoglossus** and a **chondroglossus** (see Fig. 11), as is usually the case in modern humans (see e.g., Terminologia Anatomica 1998). The ceratoglossus connects the greater horn of the hyoid bone to the tongue, while the chondroglossus connects the body of the hyoid bone (the inferior horn is usually poorly developed or absent) to the tongue. These two bundles of the hyoglossus were described in various other species of hylobatids by authors such as Kohlbrügge (1890–1892), Edgeworth (1935), Kanagasutheram (1952–1954) and Saban (1968).
- Usual innervation: Data are not available.

Styloglossus (Figs. 11, 12)
- Usual attachments: Mainly runs from the tympanic region (as described by, e.g., Kohlbrügge 1890–1892 in *H. syndactylus*, *H. agilis*, *H. moloch* and *H. lar* and corroborated in our *H. gabriellae* specimen VU HG1 in which there is no well-developed, ossified styloid process) to the tongue, its fibers running

longitudinally and passing mainly laterally to the hyoglossus; only a few fibers of the styloglossus were blended with fibers of the hyoglossus.

- Usual innervation: CN9 (Kohlbrügge 1890–1892: *H. moloch, H. agilis, H. syndactylus*).

Palatoglossus (Fig. 12)

- Usual attachments: Kohlbrügge (1890–1892) suggested that in *H. syndactylus, H. agilis, H. moloch* and *H. lar* the palatoglossus is usually not present as a distinct muscle. However, in our *H. gabriellae* specimen VU HG1 there were seemingly at least some fleshy fibers within the palatoglossal fold (on both sides of the body), which thus seem to correspond to the fibers of the palatoglossus of other taxa, connecting the soft palate to the supero-posterior portion of the tongue (and seemingly not extending to the lateral portion of the tongue) and being somewhat blended with the fibers of the styloglossus (thus supporting the idea that these two muscles derive from the same ontogenetic anlage: e.g., Diogo et al. 2008, Diogo and Wood 2011a).
- Usual innervation: Data are not available.

Sternohyoideus (1.5 g; Figs. 5, 8)

- Usual attachments: From the sternum and adjacent regions to the hyoid bone.
- Usual innervation: Exclusively by the ramus descendens of the hypoglossal nerve (Kohlbrügge 1890–1892: *H. moloch, H. agilis, H. syndactylus*).
- Notes: According to Deniker (1885) in the fetal *Hylobates* specimen dissected by him the sternohyoideus contacts its counterpart at the midline and has a tendinous intersection at about the middle of its length. Kohlbrügge (1890–1892) states that in *H. syndactylus, H. agilis,* and *H. moloch* the sternohyoideus runs from the manubrium and first and second ribs to the hyoid bone, that posteriorly the muscle lies very close to its counterpart but anteriorly they diverge and are well-separated from each other, and that there is a tendinous intersection, often in the junction of the middle and lower (posterior) third (but in *H. syndactylus* it lies in the middle of the muscle). Fig. 56 of Starck and Schneider (1960) shows a *H. syndactylus* specimen in which at least the anterior portion of the sternohyoideus seems to be well-separated from the anterior portion of its counterpart. In our *H. gabriellae* specimen VU HG1 the sternohyoideus runs from the sternum to the hyoid bone, being well-separated from its counterpart for almost its whole extent (Fig. 8); in at least one side of the body, it does seem to have a tendinous intersection.

Omohyoideus (0.8 g; Figs. 5, 8)

- Usual attachments: From the scapula to the hyoid bone, passing deep to the sternocleidomastoideus.
- Usual innervation: Exclusively by the ramus descendens of the hypoglossal nerve (Kohlbrügge 1890–1892: *H. moloch, H. agilis, H. syndactylus*).

- Notes: The intermediate tendon of the omohyoideus is missing in the hylobatid specimens reported by Deniker (1885), Sonntag (1924) and Ashton and Oxnard (1963) and examined by us; within the *H. syndactylus*, *H. agilis* and *H. moloch* specimens dissected by Kohlbrügge (1890–1892) only *H. syndactylus* had a small intermediate tendon. Ashton and Oxnard (1963) state that in the hylobatid specimens dissected by them the omohyoideus inserts onto the intermediate region of the superior border of the scapula, while in *Homo* (and, e.g., in *Pongo*) it inserts onto the lateral edge of the scapular notch. Deniker (1885) states that in the fetal *Hylobates* sp. specimen dissected by him the omohyoideus runs from the superior angle of the scapula (and not near to the coracoid process, as is usually the case in modern humans according to him) to the hyoid bone. Kohlbrügge (1890–1892) stated that in *H. agilis* the omohyoideus originates from the scapula, that in *H. syndactylus* and *H. moloch* it reaches the acromial end of the clavicle, and that in all three species it inserts onto the hyoid bone.

Sternothyroideus (0.9 g; Figs. 8, 9)

- Usual attachments: From the sternum and adjacent regions to the thyroid cartilage.
- Usual innervation: Exclusively by the ramus descendens of the hypoglossal nerve (Kohlbrügge 1890–1892: *H. moloch*, *H. agilis*, *H. syndactylus*).
- Notes: Contrary to the condition in modern humans and in most other primates, in hylobatids (e.g., Deniker 1885; Kohlbrügge 1890–1892; our study) the main body of the sternothyroideus does not usually extend anteriorly, so that its anterior portion is anterior to the posterior portion of the main body of the thyrohyoideus (Fig. 9). In the fetal *Hylobates* specimen dissected by Deniker (1885) the sternothyroideus does not contact its counterpart at the ventral midline, and has a tendinous intersection in its inferior 1/3, near its attachment to the sternum; anteriorly it is divided into two bundles, one attaching posteriorly to the thyrohyoideus and the other extending anteriorly to the posterior margin of the thyrohyoideus to insert onto the anterior and lateral border of the thyroid cartilage near the hyoid cornu. In the *H. syndactylus*, *H. agilis* and *H. moloch* specimens dissected by Kohlbrügge (1890–1892) the sternothyroideus is not in contact with its counterpart for most of its length, being well-separated from it and it has a tendinous intersection. Posteriorly it mainly originates from the sternum and anteriorly it is divided into two bundles with the most lateral of these bundles extending anteriorly to the posterior margin of the thyrohyoideus to insert onto the anterior and lateral border of the thyroid cartilage (see Fig. 1 of his Plate XI). Fig. 56 of Starck and Schneider (1960) shows a *H. syndactylus* specimen in which the sternothyroideus is divided into two bundles, with the most lateral of these bundles extending anteriorly to the posterior margin of the thyrohyoideus. In our *H. gabriellae* specimen VU HG1 the sternothyroideus runs from the sternum to the thyroid

cartilage, passing anteriorly to the posterior portion of the thyrohyoideus; the sternothyroideus is well-separated from its counterpart at the midline and from the sternohyoideus, is not fused with the thyrohyoideus and does not seem to have tendinous intersections.

Thyrohyoideus (Fig. 9)

- Usual attachments: From the thyroid cartilage to the hyoid bone.
- Usual innervation: Exclusively by the ramus descendens of the hypoglossal nerve (Kohlbrügge 1890–1892: *H. moloch, H. agilis, H. syndactylus*).
- Notes: Kohlbrügge (1890–1892) stated that in one side of the *H. agilis* specimen dissected by him some of the fibers of the pars cricopharyngea of the inferior constrictor attach onto the glandula thyroidea, forming a 'levator glandulae thyreoideae lateralis'. This could correspond to the **levator glandulae thyroideae** *sensu* the present study, but this latter muscle is seemingly a longitudinal hypobranchial muscle, and not a branchial muscle derived from the pharyngeal constrictors as described by Kohlbrügge (1890–1892).

3.5 Extra-ocular musculature

Muscles of eye (Fig. 7)

- Usual attachments: To our knowledge, there are no detailed published descriptions of these muscles in hylobatids. In the hylobatids dissected by us the **rectus inferior, rectus superior, rectus medialis, rectus lateralis, obliquus superior** and **obliquus inferior** are essentially similar to those of modern humans (Fig. 7). The **orbitalis** and **levator palpebrae superioris** are consistently found in modern humans and at least some other primate and non-primate mammals, so these muscles are very likely also present in hylobatids. Detailed studies of the eye and the extraocular muscles in hylobatids and other apes are clearly needed.
- Usual innervation: Data are not available.

CHAPTER 4

Pectoral and Upper Limb Musculature

Serratus anterior (Fig. 18)

- Usual attachments: From the ribs to the medial side of the scapula, being well-separated from the levator scapulae.
- Usual innervation: Long thoracic nerve from C5, C6 and C7 (Schück 1913b: *H. syndactylus*). Kohlbrügge (1890–1892: *H. moloch, H. agilis, H. syndactylus*) suggests it is innervated from C6 and C7 via the long thoracic nerve but also from the 'posterior thoracic nerve'—which is a synonym of the long thoracic nerve, so it is not clear what this author intended to convey.
- Notes: In hylobatids the serratus anterior can originate from ribs 1–10 (some *H. syndactylus*: Schück 1913b; *Hylobates* sp.: Hepburn 1892; some *H. lar*: Stern et al. 1980b; *H. syndactylus*: Kohlbrügge 1890–1892), ribs 1–11 (some *H. syndactylus*: Schück 1913b; *H. syndactylus* and some *H. lar*: Stern et al. 1980b; *H. agilis* and *H. moloch*: Kohlbrügge 1890–1892) or ribs 1–12 (fetal *Hylobates* sp. specimen: Deniker 1885). Stewart (1936) stated that in *H. lar* the serratus anterior is divided into two main bundles: a small superior bundle arising from ribs 1–2 and inserting onto the medial angle of the scapula in close association with the levator scapulae; a large inferior bundle arising from ribs 1–10 and inserting onto the vertebral border and inferior angle of the scapula. Andrews and Groves (1976) compiled information regarding the hylobatid specimens dissected by Bischoff (1870), Ruge (1890-1891), Kohlbrügge (1890–1892), Grönroos (1903), Plattner (1923), Kanagasuntheram (1952–1954) and Ashton and Oxnard (1963) and stated that in *H. syndactylus* the cranial and caudal heads of the serratus anterior originate respectively from ribs 1–4 and 4–10 in *H. hoolock*, from ribs 1–2 and 2–10 in *H. moloch*, from ribs 1–3 and 3–11 in *H. agilis*, from ribs 1–3 and 2–11, in *H. lar* the cranial head originates from ribs 2–9, and in *H. muelleri* the caudal head originates from ribs 2–6. Michilsens et al. (2009) stated that in the 11 hylobatid specimens dissected by them (3 *H. lar*, 2 *H. pileatus*, 2 *H. moloch* and 4 *H. syndactylus*) the serratus anterior has a main

body from ribs 2–10 to the inferior angle and medial border of the scapula, and a 'pars superior', which connects the superior angle and medial border of the scapula to ribs 2–3, except in the three *H. lar* specimens, in which this latter structure attaches to ribs 1–3.

- Synonymy: Grand dentelé (Deniker 1885); serratus anticus major (Kohlbrügge 1890–1892); serratus magnus (Hepburn 1892); pars caudalis of serratus anticus (Schück 1913b).

Rhomboideus (10.3 g; Figs. 15, 23)

- Usual attachments: From the cervical and thoracic vertebrae to the medial border of the scapula.
- Usual innervation: By C5 in *H. agilis*, *H. syndactylus* and *H. moloch* and also by C4 in *H. agilis* and *H. moloch* (Kohlbrügge 1890–1892); by C4 (in one specimen of *H. syndactylus*), by C5 (in another specimen of the same species) or by C4 and C5 (in two specimens of the same species) (Schück 1913b).
- Notes: The **rhomboideus major**, **rhomboideus minor** and **rhomboideus occipitalis** are usually not present as distinct structures in hylobatids, although both a rhomboideus major and a rhomboideus minor were reported in the fetal *Hylobates* sp. specimen dissected by Deniker (1885) and in one of the four *H. syndactylus* specimens dissected by Schück (1913b). In hylobatids the rhomboideus usually originates from C6-T4 (*Hylobates*: Sonntag 1924), T1–T6 (*H. moloch*: Kohlbrügge 1890–1892), C7-T6 (*H. lar*: Stewart 1936), C7-T7 (*H. lar*: Donisch 1973), C6-T6 (*H. agilis*: Kohlbrügge 1890–1892), C3-T7 (*H. syndactylus*: Kohlbrügge 1890–1892), or from C7 or T1 to T7 or T8 (*H. syndactylus*: Schück 1913b). Andrews and Groves (1976) compiled information regarding the hylobatid specimens dissected by Bischoff (1870) Ruge (1890-1891), Kohlbrügge (1890–1892), Grönroos (1903), Plattner (1923), Kanagasuntheram (1952–1954) and Ashton and Oxnard (1963), and stated that in *H. syndactylus* the rhomboideus originates from C2-T7, in *H. hoolock* from C6-T8, in *H. muelleri* from C6-T6, in *H. moloch* from T1–T6, and in *H. agilis* from C6-T6. Michilsens et al. (2009) stated that within the 11 hylobatids dissected by them (3 *H. lar*, 2 *H. pileatus*, 2 *H. moloch* and 4 *H. syndactylus*) the rhomboideus originated from the spinous processes of T2–T5, except in the three *H. lar* specimens, in which the origin was from T1–T5. In our *H. gabriellae* and *H. lar* specimens HU HL1 and VU HG1 the rhomboideus originates far from the occiput, in, or near to, C7 and/or T1, and forms a continuous sheet that inserts onto the whole medial border of the scapula.

Levator scapulae (Figs. 10, 17, 21, 22, 23)

- Usual attachments: From the cervical vertebrae to the anterior portion of the medial side of the scapula.
- Usual innervation: Mainly by C4 (Kohlbrügge 1890–1892: *H. moloch*, *H. agilis*, *H. syndactylus*); by C4 (in one specimen) and by C3 and C4 (in two other specimens) (Schück 1913b: *H. syndactylus*).

- Notes: In hylobatids the origin of the levator scapulae usually does not extend beyond C5. Deniker (1885: *Hylobates* sp. fetus), Plattner (1923: *H. syndactylus*), Sonntag (1924: *Hylobates*) and Michilsens et al. (2009: *H. lar, H. pileatus, H. moloch, H. syndactylus*) described an origin from C1–C4, Stewart (1936: *H. lar*) and Donisch (1973: *H. lar*) described an origin from C1–C5, Kohlbrügge (1890–1892: *H. moloch, H. agilis, H. syndactylus*) found an origin from C1, C3 and C4, and we found an origin from C1–C5 in our *H. lar* specimen HU HL1 and from C1–C3 or C1–C4 in our *H. gabriellae* specimen VU HG1, while Schück (1913b) reported an origin from C1–C3 in three *H. syndactylus* specimens.
- Synonymy: Angulaire de l'omoplate (Deniker 1885); levator anguli scapulae (Hepburn 1892, Sonntag 1924); levator scapulae plus pars cranialis of serratus anticus (Schück 1913b).

Levator claviculae (Figs. 10, 22)

- Usual attachments: From the atlas to the clavicle, near the acromial extremity of this latter bone.
- Usual innervation: Mainly by C4 in *H. agilis, H. syndactylus*, but also C3 in *H. syndactylus* (Kohlbrügge 1890–1892); in two specimens of *H. syndactylus* by C3 and C4, in another specimen by C3 only, and in a further specimen by C4 only (Schück 1913b).
- Notes: The levator claviculae was found in all hylobatid specimens dissected by us and described in the literature reviewed by us, except in one specimen of *H. moloch* reported by Kohlbrügge (1890–1892). Our observations agree with the description of Kohlbrügge (1890–1892), Schück (1913ab), Stewart (1936), Ashton and Oxnard (1963) and Andrews and Groves (1976) according to which in hylobatids the levator claviculae is deep to the trapezius. As described by Deniker (1885), Kohlbrügge (1890–1892), Chapman (1900) and Schück (1913ab), and quantitatively shown by Stewart (1936; i.e., position index from acromial end of clavicle is 18.3 in *Hylobates*, contrary to, e.g., 38.2 and 38.4 in *Gorilla* and *Pan*, respectively) and corroborated by our own dissections, in *Hylobates* the insertion of the levator claviculae on the clavicle is considerably more lateral than in other non-human hominoids: it usually lies near the acromial extremity of the clavicle or even extends to the acromioclavicular joint. The **atlantoscapularis posticus** (see, e.g., Diogo et al. 2009a) is usually not present as a distinct muscle in hylobatids.
- Synonymy: Cleido-omotransversarius, omocervicalis, cleido-cervicalis, acromio-cervicalis or levator anticus scapulae (Deniker 1885, Barnard 1875, Kohlbrügge 1890–1892, Schück 1913b); acromiotrachelian, clavotrachelian, cleidoatlanticus or cleido-omotransversale (Miller 1932); atlantoscapularis (Stewart 1936); atlantoscapularis anterior (Ashton and Oxnard 1963).

Subclavius (1.8 g; Figs. 18, 21, 25)

- Usual attachments: From the first rib and often also from rib 2 and/or rib 3 to the clavicle.

- Usual innervation: Nerve to subclavius, from C6 (Hepburn 1892: *Hylobates* sp.; Kohlbrügge 1890–1892: *H. moloch H. agilis, H. syndactylus*).
- Notes: In hylobatids the origin of the subclavius often extends to rib 3 and/or its costal cartilage, as described by Kohlbrügge (1890–1892), Hepburn (1892), Sonntag (1924), Miller (1932), Andrews and Groves (1976) and Michilsens et al. (2009) and corroborated by our own dissections of the *H. lar* specimen HU HL1. It should be noted that Michilsens et al. (2009) only reported such an origin from rib 3 in the three *H. lar* specimens dissected by them, i.e., they did not report an origin from this rib in the two *H. pileatus*, the two *H. moloch*, and the four *H. syndactylus* that they analyzed. Also, in the fetal *Hylobates* sp. specimen dissected by Deniker (1885) and in our *H. gabriellae* specimen VU HG1 there was no origin from rib 3. However, Andrews and Groves (1976) reviewed information regarding the hylobatid specimens dissected by Bischoff (1870), Ruge (1890-1891), Kohlbrügge (1890–1892), Grönroos (1903), Plattner (1923), Kanagasutheram (1952–1954) and Ashton and Oxnard (1963), and concluded that in *H. syndactylus* the muscle usually originates from ribs 2–3, in *H. hoolock* from ribs 1–2, in *H. muelleri* from ribs 2–3, in *H. moloch* from ribs 2–3 and in *H. agilis* from ribs 2–3, thus indicating that such an origin from rib 3 also often occurs in members of various hylobatid species other than *H. lar*. The **costocoracoideus** is not present as a distinct muscle in hylobatids, but these primates do usually have a ligamentum costocoracoideum (which, according to Deniker 1885, is deeply blended with the subclavius, thus supporting the hypothesis that this ligament corresponds to the muscle costocoracoideus of mammals such as monotremes: see Diogo et al. 2009a).
- Synonymy: Sous-clavier (Deniker 1885).

Pectoralis major (Figs. 19, 20, 25)
- Usual attachments: The pars clavicularis runs from the medial portion of the clavicle and anterior portion of the sternum to the proximal portion of the humerus; the pars sternocostalis runs from the sternum and ribs to the proximal portion of the humerus, inserting proximally to the insertion of the pars clavicularis; the pars abdominalis runs from the ribs and aponeurosis of the external oblique to the proximal portion of the humerus and the coracoid process of the scapula, being deep to the pars sternocostalis and pars clavicularis.
- Usual innervation: Anterior thoracic (pectoral) nerves, from C5 and C6 in *H. moloch, H. agilis,* and *H. syndactylus*, and also from C7 in *H. agilis* (Kohlbrügge 1890–1892); lateral and medial pectoral nerves (Hepburn 1892: *Hylobates* sp.).
- Notes: In hylobatids (e.g., Kohlbrügge 1890–1892; Ruge 1890-1891; Miller 1932; Howell and Straus 1932; Andrews and Groves 1976; Jungers and Stern 1981; Payne 2001; Michilsens et al. 2009; our dissections) the abdominal head

of the pectoralis major is usually blended with the biceps brachii. According to Loth (1931) in 100% of hylobatids the pectoral major is completely separated from its counterpart at the midline. Andrews and Groves (1976) compiled information regarding the hylobatid specimens dissected by Bischoff (1870), Ruge (1890-1891), Kohlbrügge (1890–1892), Grönroos (1903), Plattner (1923), Kanagasuntheram (1952–1954) and Ashton and Oxnard (1963) and stated that: in *H. syndactylus* the clavicular origin of the pectoralis major is from the medial 1/2 of the clavicle, the abdominal portion of this muscle originates from ribs 5–6 and the muscle does not reach the midline; in *H. hoolock* the clavicular origin of the pectoralis major is from the medial 3/4 of the clavicle, the abdominal portion originates from rib 5 and the muscle does reach the midline; in *H. muelleri* the clavicular origin is from the whole area of the clavicle, the abdominal portion originates from rib 6 and the muscle does reach the midline; in *H. moloch* the clavicular origin is from the medial 1/4 of the clavicle, the abdominal portion originates from ribs 4–6 and the muscle does not reach the midline; in *H. agilis* the clavicular origin is from the medial 1/2 of the clavicle, the abdominal portion originates from rib 6, and the muscle does not reach the midline; in *H. lar* the abdominal portion of this muscle originates from rib 7. Authors such as Sonntag (1924) and Miller (1932) reported an independent 'pectoralis abdominis' in hylobatids, but this structure very likely corresponds to the pars abdominalis of the pectoralis major *sensu* the present study. Van den Broek (1909) described a **sternalis** and a 'pectoralis quartus' (both as an anomaly) in a specimen of *H. syndactylus*, the '**pectoralis quartus**' running from the abdominal head of the pectoralis major and the sheat of the rectus abdominis to the humerus (so, in this case, this 'pectoralis quartus' is effectively a distinct, anomalous structure, i.e., it does not correspond to the pars abdominalis, nor to the pars clavicularis or sternocostalis, of the pectoralis major *sensu* the present study). Stern et al. (1980a) stated that EMG in *H. lar* revealed that although the 'cranial portion' of the pectoralis major (which includes the pars clavicularis plus a portion of the pars sternocostalis *sensu* the present study) assists the 'caudal portion' of this muscle during retraction of the protracted (elevated) forelimb, the unique role of the 'cranial portion' being the flexion of the adducted forelimb as required in the recovery phase of the locomotor cycle.

- Synonymy: Pectoralis major plus lower portion of pectoralis minor (Hartmann 1886); pectoralis major plus pectoralis abdominis, abdominalis, and/or chondroepitrochlearis quartus (Sonntag 1924b, Miller 1932, Aiello and Dean 1990, Gibbs 1999).

Pectoralis minor (Figs. 18, 20, 25)
- Usual attachments: From the ribs to the coracoid process of the scapula and, often, to the clavicle.

- Usual innervation: Anterior thoracic (pectoral) nerves, from C5 and C6, in *H. moloch*, *H. agilis* and *H. syndactylus*, and also from C7 in *H. agilis* (Kohlbrügge 1890–1892); medial pectoral nerve (Hepburn 1892: *Hylobates* sp.).
- Notes: In hylobatids the pectoralis minor is often at least partially inserted onto the clavicle; this was corroborated by Hepburn (1892), Sonntag (1924), Stewart (1936), Gibbs (1999), Michilsens et al. (2009) and by our dissections, although such an insertion onto the clavicle was not reported in the fetal *Hylobates* sp. specimen dissected by Deniker (1885) and in the hylobatid specimens examined by Kohlbrügge (1890–1892). The pectoralis minor originates from ribs 2–3 (our *H. gabriellae* specimen VU HG1), 2–4 (*H. pileatus*, *H. moloch*, *H. syndactylus*: Michilsens et al. 2009; our *H. lar* specimen HU HL1, in which the muscle and originates from the intercostal space between ribs 4–5, but without reaching rib 5), 1–5 (*Hylobates* sp.: Hepburn 1892; *H. lar*: Michilsens et al. 2009), 3–5 (*H. lar*: Stewart 1936; *Hylobates*: Sonntag 1924), or 4–5 (fetal *Hylobates* sp. specimen: Deniker 1885). Andrews and Groves (1976) compiled information regarding the hylobatid specimens dissected by Bischoff (1870), Ruge (1890-1891), Kohlbrügge (1890–1892), Grönroos (1903), Plattner (1923), Kanagasuntheram (1952–1954) and Ashton and Oxnard (1963) and stated that in *H. syndactylus* the muscle usually originates from ribs 3–4, in *H. hoolock* from ribs 3–5, in *H. muelleri* from ribs 6–7, in *H. moloch* from ribs 4–5, in *H. agilis* from ribs 2–5, and in *H. lar* from rib 6. The **pectoralis tertius ('xiphihumeralis')**, **sternalis**, and **panniculus carnosus** (see, e.g., Diogo et al. 2009a) are usually not present as distinct muscles in hylobatids, although Van den Broek (1909) and Jouffroy (1971) stated that the sternalis might be occasionally found in these primates.
- Synonymy: Upper portion of pectoralis minor (Hartmann 1886).

Infraspinatus (13.3 g; Figs. 15, 23, 24, 26)
- Usual attachments: From the infraspinous fossa of the scapula and the infraspinatus fascia to the greater tuberosity of humerus and in many cases to the capsule of the glenohumeral joint.
- Usual innervation: Suprascapular nerve, from C5 (Kohlbrügge 1890–1892: *H. moloch*, *H. agilis*, *H. syndactylus*).

Supraspinatus (10.7 g; Figs. 15, 23, 24, 26)
- Usual attachments: From the supraspinous fossa of the scapula and the supraspinatus fascia to the greater tuberosity of humerus and in many cases to the capsule of the glenohumeral joint.
- Usual innervation: Suprascapular nerve from C5 (*H. moloch*, *H. agilis*, *H. syndactylus*).
- Notes: The muscle **scapuloclavicularis**, occasionally present in modern humans, has not been described in hylobatids nor was it found in the hylobatids dissected by us.

Deltoideus (48.1 g; Figs. 18, 19, 20, 21, 22, 25)
- Usual attachments: From the lateral portion of the clavicle (pars clavicularis), the acromion (pars acromialis) and the spine of the scapula and the infraspinatus fascia (pars spinalis) to the humerus.
- Usual innervation: Axillary nerve from C5 and C6 (Kohlbrügge 1890–1892: *H. moloch, H. agilis, H. syndactylus*); axillary nerve (Hepburn 1892: *Hylobates* sp.).
- Notes: In hylobatids the insertion of the deltoideus onto the humeral shaft is usually more distal than in other hominoids, the muscle often extending to the middle of this bone; regarding the origin from the clavicle, it is often from the lateral 1/3 or 1/4 of this bone (e.g., Hepburn 1892, Andrews and Groves 1976, Michilsens et al. 2009).

Teres minor (4.8 g; Figs. 24, 26)
- Usual attachments: From the infraspinatus fascia and the lateral border of the scapula to the greater tuberosity of humerus, occasionally also extending to the shaft distal to the tuberosity.
- Usual innervation: Axillary nerve from C5 and C6 (Kohlbrügge 1890–1892: *H. moloch, H. agilis, H. syndactylus*); axillary nerve (Hepburn 1892: *Hylobates* sp.).
- Notes: An insertion of the teres minor onto the shaft of the humerus, distal to the greater tuberosity, was not reported by most authors nor found in our *H. lar* specimen HU HL1, but it was found in our *H. gabriellae* specimens VU HG1 and VU HG2. As described by authors such as Ashton and Oxnard (1963) and corroborated by our dissections, in some hylobatids the teres minor is blended on its deep aspect with the infraspinatus.
- Synonymy: Petit rond (Deniker 1885).

Subscapularis (23.7 g; Figs. 18, 21, 25, 26)
- Usual attachments: From the subscapular fossa of the scapula to the lesser tuberosity of the humerus.
- Usual innervation: Subscapular nerves from C5 and C6 (Kohlbrügge 1890–1892: *H. moloch, H. agilis, H. syndactylus*); subscapular nerves (Hepburn 1892: *Hylobates* sp.).
- Notes: To our knowledge an insertion of the subscapularis onto the shaft of the humerus distal to the lesser tuberosity has never been reported in hylobatids; it is also not found in the hylobatid specimens dissected by us, except possibly in the *H. gabriellae* specimen VU HG2 (in which a few fibers seem to extend distally to the tuberosity). As noted by Kohlbrügge (1890–1892), Kanagasutheram(1952–1954) and Andrews and Groves (1976) and corroborated by our dissections, hylobatids usually have a distinct, peculiar pars inferioris of the subscapularis,

which is partially separated, medially, from the main, anterior portion of the muscle by a ridge of the scapula (see, e.g., Fig. 21).

Teres major (15.8 g; Figs. 21, 24, 25, 26)
- Usual attachments: From the lateral border and inferior angle of the scapula to the proximal portion of the humerus by means of a distal tendon that is usually at least partially fused to the distal tendon of the latissimus dorsi.
- Usual innervation: Subscapular nerves from C7 and C8 (Kohlbrügge 1890–1892: *H. moloch, H. agilis, H. syndactylus*); subscapular nerves (Hepburn 1892: *Hylobates* sp.).
- Notes: In hylobatids the distal tendons of the latissimus dorsi and of the teres major are usually partially or completely fused to each other at their insertions onto the humerus (corroborated by, e.g., Hepburn 1892; Kohlbrügge 1890–1892; Miller 1932; Stewart 1936; Michilsens et al. 2009; also corroborated by our dissections, although in an adult *H. lar* specimen recently dissected by S. Dunlap the tendons seemed to be somewhat separated: pers. comm.). According to Gibbs (1999) in hylobatids the origin from the scapula is from more than 1/2 of the lateral border of this bone. Kohlbrügge (1890–1892) stated that in *H. moloch, H. agilis,* and *H. syndactylus* the teres majors originates from 1/2 of the lateral border of the scapula. Michilsens et al. (2009) stated that in the 11 hylobatid specimens dissected by them (3 *H. lar*, 2 *H. pileatus*, 2 *H. moloch* and 4 *H. syndactylus*) the teres major originates from the inferior angle of the scapula, except in the three *H. lar* specimens in which the muscle also originates from the lower third of the lateral border of the scapula. In our *H. lar* specimen HU HL1 the teres major originates from the medial 2/3 (6 cm in a total of 9 cm) of the lateral border of the scapula, being blended with the teres minor laterally and the subscapularis medially; it inserts onto the shaft of the humerus distal to the lesser tuberosity of the humerus. In our *H. gabriellae* specimen VU HG1 the teres major runs from the medial 1.6 cm (of the total 4.3 cm) of the lateral border of the scapula to the proximal portion of the humerus.
- Synonymy: Grand rond (Deniker 1885).

Latissimus dorsi (Figs. 15, 18, 22, 23, 24, 25, 26)
- Usual attachments: From the vertebrae, ribs, thoracolumbar fascia and, often, directly and/or indirectly from the pelvis, to the proximal humerus.
- Usual innervation: Subscapular nerves from C7 and C8 (Kohlbrügge 1890–1892: *H. moloch, H. agilis, H. syndactylus*); thoracodorsal (middle scapular) nerve (Hepburn 1892: *Hylobates* sp.).
- Notes: Andrews and Groves (1976) stated (based on Bischoff 1870, Ruge 1890–1891, Kohlbrügge 1890–1892, Hepburn 1892, Primrose 1899, 1900, Grönroos 1903, Plattner 1923, Sonntag 1924b, Sullivan and Osgood 1927, Howell and Straus 1931, Stewart 1936, Kanagasuntheram 1952–1954 and Ashton and Oxnard 1963) that in hylobatids the latissimus dorsi has a 'cranial' origin from

T6 or T7, a 'caudal' origin from the iliac crest and an extensive costal origin and that in *H. syndactylus* the fleshy slips of latissimus dorsi originate from ribs 7–13, the 'caudal' origin being fleshy from T9–10 and the aponeurosis from the iliac crest extending up to T10, in *H. hoolock* the fleshy slips originate from ribs 7–13, the 'caudal' origin being aponeurotic from T8–13 and the aponeurosis from the iliac crest extending up to all vertebrae, in *H. moloch* the fleshy slips originate from ribs 8–13, the caudal origin being tendinous from T8–9 and the aponeurosis from the iliac crest extending up to T10, while in *H. agilis* the fleshy slips originate from ribs 8–13, the caudal origin being fleshy from T8–9 and the aponeurosis from the iliac crest extending up to T12. Michilsens et al. (2009) stated that in the hylobatid specimens dissected by them (3 *H. lar*, 2 *H. pileatus*, 2 *H. moloch* and 4 *H. syndactylus*) the latissimus dorsi runs from the lower 6 thoracic vertebrae, the iliac crest and the lower 4 ribs to the bicipital groove of the humerus, except in the three *H. lar* specimens, in which the muscle does not always originate from the iliac crest. Donisch (1973) stated that in the *H. lar* female he dissected the latissimus dorsi originates from ribs 8–13 and spinous processes of T7–T13 (i.e., there was no origin from the iliac crest).

- Synonymy: Grand dorsal (Deniker 1885).

Dorsoepitrochlearis (Figs. 18, 21, 22, 24, 25, 26)

- Usual attachments: From the distal portion of the latissimus dorsi to the medial epicondyle of the humerus and occasionally also to the adjacent shaft of the humerus.
- Usual innervation: Radial nerve (Kohlbrügge 1890–1892: *H. moloch*, *H. agilis*, *H. syndactylus*; Hepburn 1892: *Hylobates* sp.).
- Notes: Within hylobatids Barnard (1875), Kohlbrügge (1890–1892), Chapman (1900), Schück (1913a), Miller (1932), Ashton and Oxnard (1963), Andrews and Groves (1976), Jungers and Stern (1981) and Michilsens et al. (2009) found a bony insertion onto the medial epicondyle of the humerus, as we did, while Payne (2001) refers to an insertion onto the distal humerus just proximal to the medial epicondyle; on one side of the fetal *Hylobates* sp. specimen dissected by Deniker (1885) there was an insertion onto the aponeurosis of the arm only, as found in *H. moloch* by Bischoff (1870), but on the other side of that fetus the insertion extended to the medial epicondyle of the humerus. In hylobatids the dorsoepitrochlearis is usually deeply blended with the short head of the biceps brachii, as noted by Howell and Straus (1932), Andrews and Groves (1976), Jungers and Stern (1980, 1981) and Michilsens et al. (2009) and corroborated by our dissections. In our *H. lar* and *H. gabriellae* specimens HU HL1 and VU HG1 the dorsoepitrochlearis originates mostly from the latissimus dorsi, although it is also indirectly associated with the teres major because the tendon of this latter muscle is blended with the tendon of the

latissimus dorsi; as described by Howell and Straus (1932; see, e.g., their Plate 1) the dorsoepitrochlearis is deeply blended with the short head of the biceps, some of its fibers attaching directly (as described by e.g., Barnard 1875) onto the distal 1/2 of the humerus as well as onto the muscular septum between the long head of the triceps brachii and the brachialis (the dorsoepitrochlearis thus also being somewhat blended with both these muscles) and onto the medial epicondyle of the humerus, but most of its fibers attach together with the short head of the biceps brachii on the flexor digitorum superficialis and also on the flexor carpi radialis.

- Synonymy: Latissimo-condylus or latissimo-epitrochlearis (Barnard 1875); latissimo-condyloideus (Kohlbrügge 1890–1892, Hepburn 1892, Chapman 1900, Grönroos 1903); latissimo-tricipitalis (Schück 1913ab).

Triceps brachii (40.0 g; Figs. 21, 22, 24, 26)
- Usual attachments: From the lateral portion of the lateral border of scapula (caput longum) and the shaft of the humerus (caput laterale and caput mediale) to the olecranon process of the ulna.
- Usual innervation: Radial nerve (Kohlbrügge 1890–1892: *H. moloch, H. agilis, H. syndactylus*; Hepburn 1892: *Hylobates* sp.).
- Notes: Within hylobatids Kohlbrügge (1890–1892) and Loth (1931) reported an origin of the long head from the lateral 1/3 of the lateral border of the scapula and this was corroborated in our dissections; Gibbs (1999) reported 1/3 to 1/2 in hylobatids, so the usual condition for these primates seems to be that in which the long head originates from less than 1/2 of the lateral border of the scapula. In our *H. lar* specimen HU HL1 the lateral head originates from the lateral margin of the proximal region of the humerus (not from intermuscular septum), the long head from the lateral 3.5 cm (of the total 9 cm) of the lateral border of the scapula, and the medial head from the medial part of the distal 1/2 of the humerus; the three heads fuse (at the 1/2 distal of the humerus), and give rise to a strong and broad tendon that inserts onto the olecranon process of the ulna. In our *H. gabriellae* specimen VU HG1 the long head originates from the lateral 1.5 cm (of the total 4.3 cm) of the lateral border of the scapula, the medial head from the distal 1/2 to 1/3 of the humerus, and the lateral head from the region just below (distal to) the anatomical neck of the humerus; the three heads fuse and insert, through a single tendon, onto the olecranon process of the ulna.
- Synonymy: Multiceps extensor cubiti (Barnard 1875); anconeus longus, externus and internus (Kohlbrügge 1890–1892); triceps extensor cubiti (Hepburn 1892).

Brachialis (33.5 g; Figs. 24, 27)
- Usual attachments: From the humeral shaft (well distal to the surgical neck of the humerus) to the ulnar tuberosity.

- Usual innervation: 'Median nerves', but these nerves seem to also include the musculocutaneous nerve as they also supply the skin (Kohlbrügge 1890–1892: *H. moloch*, *H. agilis*, *H. syndactylus*); musculocutaneous nerve (Hepburn 1892, Koizumi and Sakai 1995: *Hylobates* sp.); from a common trunk representing the united musculocutaneous, median and ulnar nerves (Bolk 1902: *Hylobates* sp.).
- Notes: In our *H. lar* specimen HU HL1 the brachialis runs from the distal 1/2 of the humeral shaft (from about 12 cm of the 23 cm total length of the humerus) to the ulnar tuberosity; proximally it is mainly continuous with the deltoideus, but it does not contact (i.e., it is distal to) the coracobrachialis, whereas it is somewhat blended with the dorsoepitrochlearis and the triceps brachii, while distally it is deeply blended with the brachioradialis. In our *H. gabriellae* specimen VU HG1 the brachialis is not completely divided into two heads, and originates from the humeral shaft, extending a few mm proximally to the midpoint of the humerus, inserting onto the ulnar tuberosity.
- Synonymy: Brachialis anticus, brachialis internus or brachialis anterior (Deniker 1885, Kohlbrügge 1890–1892, Hepburn 1892, Chapman 1900).

Biceps brachii (48.4 g; Figs. 18, 19, 20, 21, 24, 25, 26, 27)
- Usual attachments: From the proximal humerus (caput breve) and the supraglenoid tubercle of the scapula (caput longum), to the bicipital tubercle of the radius (common tendon) and to fascia covering the forearm flexors (bicipital aponeurosis, which in hylobatids is usually at least partially fleshy, forming a 'lacertus carnosus').
- Usual innervation: 'Median nerves' but these nerves seem to also include the musculocutaneous nerve, as they also supply the skin (Kohlbrügge 1890–1892: *H. moloch*, *H. agilis*, *H. syndactylus*); musculocutaneous nerve (Hepburn 1892, Koizumi and Sakai 1995: *Hylobates* sp.); from a common trunk representing the united musculocutaneous, median and ulnar nerves (Bolk 1902: *Hylobates* sp.).
- Notes: As described by Kohlbrügge (1890–1892), Howell and Straus (1932), Andrews and Groves (1976), Jungers and Stern (1981), Gibbs (1999) and Michilsens et al. (2009) and corroborated by our dissections, in hylobatids the distal portion of the biceps brachii is deeply blended with the proximal portion of the flexor digitorum superficialis. As described by Kohlbrügge (1890–1892), Loth (1931) and Jouffroy (1971) and corroborated by our dissections, in hylobatids the bicipital aponeurosis is usually at least partially fleshy, thus forming a 'lacertus carnosus' between the main body of the biceps brachii and the flexor muscles of the forearm. As noted by Owen (1868), Bischoff (1870), Kohlbrügge (1890–1892), Hepburn (1892), Chapman (1900), Sonntag (1924), Howell and Straus (1932), Miller (1932), Jouffroy (1971), Andrews and Groves (1976), Jungers and Stern (1981), Gibbs (1999) and Michilsens

et al. (2009) and corroborated in our *H. lar* specimen HU HL1 and in our *H. gabriellae* specimen VU HG2, within hylobatids the short head of the biceps brachii is usually at least partially originates from the humerus, although in our *H. gabriellae* specimen VU HG1 this head is exclusively originates from the coracoid process. Howell and Straus (1932) stated that in both sides of the *H. lar* specimen dissected by them the long head of the biceps brachii inserts onto the radius; these authors opted to name the other head as the 'humeral' head and not as the 'short head' because "it differed markedly from the latter and there is no absolute assurance that it was derived from the more usual breve", originating from the lesser tuberosity of the humerus and not from the coracoid process. The pectoralis major inserts directly upon the tendon of origin of this 'humeral' head; the dorsomedial border of the proximal third of the fleshy portion of this humeral head fuses with the dorsoepitrochlearis and the medial intermuscular septum and a slender tendon extending along the dorsomedial border of this head fuses with the other slender tendon of the dorsoepitrochlearis. According to Howell and Straus (1932) all these direct connections between the biceps brachii and the pectoralis major, the flexor digitorum superficialis and the dorsoepitrochlearis (and thus indirectly the latissimus dorsi), which connect the axial skeleton directly to the phalanges of the digits (e.g., origin of pectoralis major to insertion of flexor digitorum superficialis) are what we would now refer to as apomorphic features related to the peculiar brachiating adaptations of hylobatids. The two heads of the biceps brachii fuse at about the middle of the brachium, but the 'humeral' head again separates to insert onto the flexor digitorum superficialis, suggesting that the 'humeral' head of the biceps brachii and the flexor digitorum superficialis "may act as one continuous, long muscle". Kohlbrügge (1890–1892) found in *H. moloch* a long head, a short head and a 'humeral head' of the biceps brachii, suggesting an incomplete migration of the insertion of the long head from the coracoid to the humerus; according to this author in *H. syndactylus* the coracoid origin of the biceps brachii was 'vestigial', and in *H. agilis* it was entirely absent, i.e., there was only a long head and a 'humeral head'. The 'humeral head' *sensu* Kohlbrügge (1890–1892) and Howell and Straus (1932) thus clearly derives from (i.e., corresponds to part or the totality of) the short head of other primates. In our *H. lar* specimen HU HL1 the biceps brachii has two heads, the short head originating mostly from the lesser tuberosity of the humerus and the long head originating from the glenohumeral joint. At about halfway along the proximo-distal length of the humerus the two heads become deeply blended and they also blend (particularly the short head) with the dorsoepitrochlearis. More distally near the elbow the long head is somewhat separated from the short head plus the dorsoepitrochlearis, inserting mainly on the radius, while the short head plus the dorsoepitrochlearis insert distally mainly on the proximal portions of the flexor carpi radialis and particularly of the flexor digitorum superficialis although some fibers of the dorsoepitrochlearis insert distally onto the medial epicondyle of the humerus. As explained above, in our *H. gabriellae* specimen VU HG1 the long head originates from the supraglenoid tubercle of the scapula and the short originates from the coracoid process; the two heads meet and give rise to a tendon going to the radius and to a 'lacertus carnosus' going to the forearm flexors.

- Synonymy: Biceps flexor cubiti (Owen 1868, Hepburn 1892, Sonntag 1924).

Coracobrachialis (5.4 g; Figs. 20, 21, 26)
- Usual attachments: From the coracoid process of the scapula to the proximal end of the humerus.
- Usual innervation: 'Median nerves', but these nerves seem to also include the musculocutaneous nerve, as they also supply the skin (Kohlbrügge 1890–1892: *H. moloch, H. agilis, H. syndactylus*); musculocutaneous nerve (Hepburn 1892, Koizumi and Sakai 1995: *Hylobates* sp.); from a common trunk representing the united musculocutaneous, median and ulnar nerves (Bolk 1902: *Hylobates* sp.).
- Notes: Parsons (1898ab) stated that hylobatids have a distinct coracobrachialis profundus, but the observations of most authors, as well as our dissections, clearly show that the usual condition for these primates is that the **coracobrachialis superficialis/coracobrachialis longus** and **coracobrachialis profundus/coracobrachialis brevis** are not present as distinct structures. Contrary to other hominoids and to most other primates, in hylobatids (and also in gorillas) the usual condition seems to be that in which the musculocutaneous nerve does not pass through the coracobrachialis, as noted by Hepburn (1892), Howell and Straus (1932) and Gibbs (1999) and as also found in our dissections. In our *H. lar* specimen HU HL1 the coracobrachialis has a single bundle (which corresponds to the **coracobrachialis medius/coracobrachialis proprius** of other mammals) that is not pierced by the musculocutaneous nerve and that runs from the coronoid process of the scapula (not from the intermuscular septum) to the proximal 1/4 of the humerus (to about 6 cm in a total humeral length of 23 cm); in our *H. gabriellae* specimen VU HG1 the coracobrachialis has a single head, from the coracoid process to the humeral shaft and does not seem to be perforated by the musculocutaneous nerve.

Pronator quadratus (1.4 g; Fig. 36)
- Usual attachments: From the distal portion of the ulna to the distal portion of radius.
- Usual innervation: Branch/division of median nerve (Kohlbrügge 1890–1892: *H. moloch, H. agilis, H. syndactylus*); posterior interosseous nerve (Hepburn 1892: *Hylobates* sp.).
- Notes: Hepburn (1892) stated that in *Hylobates* sp. the pronator quadratus is in general more oblique than in modern humans. Deniker (1885) reported a fetal

Hylobates sp. specimen in which the muscle is markedly oblique, the insertion onto the ulna (proximo-distally) being two times longer than the insertion onto the radius. Contrary to Deniker (1885), Kohlbrügge (1890–1892) stated that in *H. moloch*, *H. agilis* and *H. syndactylus* the proximodistal length of origin on the ulna is similar to that of the insertion onto the radius and Michilsens et al. (2009) stated that in the 11 hylobatids dissected by them (3 *H. lar*, 2 *H. pileatus*, 2 *H. moloch* and 4 *H. syndactylus*) the pronator quadratus runs from the distal 1/4 of the ulna to the distal 1/4 of radius. In our *H. lar* specimen HU HL1 the muscle is more oblique than in most modern humans (its most proximal attachment onto the ulna is about 1.5 or 2 cm distal to its more proximal attachment onto the radius). This is also the case in our *H. gabriellae* specimen VU HG1 (the proximodistal length of its attachment onto the ulna being 1.7 cm, that onto the radius being 1.2 cm).

- Synonymy: Carré pronateur (Deniker 1885).

Flexor digitorum profundus (flexor digitorum profundus + flexor pollicis longus = 35.1 g; Figs. 27, 29, 31, 34, 35)

- Usual attachments: From the radius, ulna, and medial epicondyle of the humerus and interosseous membrane to distal phalanges of digits 2, 3, 4 and 5.
- Usual innervation: Median nerve to digits 2 and 3 and ulnar nerve to digits 4 and 5 (Kohlbrügge 1890–1892: *H. moloch*, *H. agilis*, *H. syndactylus*; Hepburn 1892: *Hylobates* sp.).
- Notes: Among the primates dissected by us, hylobatids and modern humans are the only ones in which the flexor pollicis longus is usually present as a distinct, independent muscle. Our dissections and the reports of most authors (e.g., Deniker 1885; Hartmann 1886; Kohlbrügge 1890–1892; Hepburn 1892; Keith 1894b; Chapman 1900; McMurrich 1903a,b; Sonntag 1924; Howell 1936a,b; Straus 1942a,b; Jouffroy and Lessertisseur 1960; Tuttle 1969; Jouffroy 1971; Van Horn 1972; Lorenz 1974; Susman 1994, 1998; Stout 2000, Tocheri et al. 2008) clearly indicate that hylobatids usually have an independent flexor pollicis longus going exclusively to digit 1; a few authors (e.g., Bischoff 1870; Payne 2001) state that in the hylobatids dissected by them the flexor pollicis longus is not really separate from the flexor digitorum profundus. In the hylobatids dissected by us we found an origin of the flexor digitorum profundus from the medial epicondyle of the humerus, ulna, radius and often the interosseous membrane and this was also the case in most hylobatids described by Kohlbrügge (1890–1892), Hepburn (1892), Tuttle (1969) and Michilsens et al. (2009), although Deniker (1885) did not describe an origin from the humerus in the *Hylobates* sp. fetus dissected by him. In our *H. lar* specimen HU HL1 the flexor digitorum profundus runs from the ulna, radius, interosseous membrane and medial epicondyle of the humerus to the distal phalanges of digits 2–5; the

flexor pollicis longus is a distinct muscle running from the radius, interosseous membrane and flexor digitorum profundus to the distal phalanx of digit 1. On both sides of our *H. gabriellae* specimen VU HG1 the flexor digitorum profundus originates from the ulna, the medial epicondyle of the humerus, the interosseous membrane and the radius and gives rise to 4 tendons, one, thin, to the distal phalanx of digit 2, and the other three to the distal phalanges of digits 3, 4 and 5; the flexor pollicis longus originates from the radius and interosseous membrane and gives rise to two tendons, one, thin, to the distal phalanx of digit 2, and the other to the distal phalanx of the thumb (digit 1) (Fig. 35). That is, in at least this latter specimen we see a 'transition' from the configuration seen in various primates, i.e., in which the radial belly of the flexor digitorum profundus goes to digits 1 and 2, to the configuration usually found in various other hylobatids and in modern humans in which there is a distinct flexor pollicis longus going only to digit 1, possibly due to the division of the radial belly of the flexor digitorum profundus into two structures and/or to the migration of its belly to digit 2 towards the other bellies (the radial belly then having a single tendon, to digit 1, and thus forming the flexor pollicis longus *sensu* the present study). Tuttle (1969) stated that the partial origin of the flexor digitorum profundus from the humerus seen in hylobatids is probably related to the importance of flexion at the elbow to help propel the body forward while the hand is fixed on a support. According to this author the prime flexor of the distal phalanx of digit 1 in hylobatids is the 'radial component of the flexor digitorum profundus musculature' (flexor pollicis longus *sensu* the present study), which is independent from the ulnar and humeral components of the flexor digitorum profundus, except for a variably developed connective tissue 'shunt' at the level of the carpus. The 'flexor shunt' is commonly less dense than the tendons that it connects; frequently it appears as a slender flat (4–5 mm wide) structure that passes obliquely inferomedially from tendon I to the lateral edge of the common tendon of digits 2–5. The 'flexor shunt' probably subserves different functions during the various major activities of the hylobatids. For instance, when the hand is used as an anatomical hook during vigorous arm-swinging, the 'shunt' probably transfers some of the contractile force of the 'radial flexor digitorum profundus' musculature to the common deep flexor tendon and thereby assists to stabilize the wrist. By contrast, when the hand is used for fine manipulation, particularly with the metacarpophalangeal joints of digits 2–5 flexed, the 'shunt' is probably somewhat slack, allowing the deep radial musculature to act as an independent flexor of the distal phalanx of the thumb. Lastly, when the thumb is widely abducted to grasp large branches, the 'shunt' is probably again stretched to the extent that the distal phalanges of the thumb and the medial four digits are flexed synchronously to produce secure grips (see Fig. 17 of Tuttle 1969). As a result, contrary to the condition in *Pan*, *Pongo* and *Gorilla*, the pollex normally plays an active role in maintaining hylobatids in suspended postures and locomotion. Based on observations of living primates Van Horn (1972) suggested that the flexor pollicis longus of hylobatids is important during the arm-pull phase of climbing, in which the terminal phalanx of the thumb is flexed as the animal lifts its weight from a previous support, i.e., this muscle supplies an important component of the power used in grasping. Stout (2000) found that the tendon of the flexor pollicis longus of hylobatids does not flex the distal phalanx of the thumb: rather it stabilizes the pollical interphalangeal joint, adducts the thumb, and flexes the pollical metacarpophalangeal joint. Susman (1998) stated that when he attempted to stimulate the flexor pollicis longus in a *H. lar* specimen with a indwelling electrode he could not elicit flexion of the pollical distal phalanx without flexion of the digit 2 as well (but later, when the animal was dead, dissection revealed that the fibers of the flexor pollicis longus were not entirely separated from those of the flexor digitorum profundus).

- Synonymy: Flexor digitorum communis profundus (Barnard 1875); fléchisseur profond (Deniker 1885); flexor profundus digitorum (Chapman 1900).

Flexor pollicis longus (flexor digitorum profundus + flexor pollicis longus = 35.1 g; Figs. 29, 31, 35, 36)

- Usual attachments: From the radius, the interosseous membrane and, sometimes, also from the flexor digitorum profundus, to the distal phalanx of digit 1 and occasionally also to distal phalanx of digit 2 (see notes about flexor digitorum profundus, above).
- Usual innervation: Branch/division of median nerve (Kohlbrügge 1890–1892: *H. moloch, H. agilis, H. syndactylus*; Hepburn 1892: *Hylobates* sp.).
- Notes: See flexor digitorum profundus above.
- Synonymy: Flexor longus pollicis (Chapman 1900); radial component of the flexor digitorum profundus (Tuttle 1969).

Flexor digitorum superficialis (35.5 g; Figs. 27, 28, 29, 31, 34)

- Usual attachments: From the radius (caput radiale), the ulna and the medial epicondyle of the humerus (caput humeroulnare) to the middle phalanges of digits 2–5.
- Usual innervation: Median nerve (Kohlbrügge 1890–1892: *H. moloch, H. agilis, H. syndactylus*; Hepburn 1892: *Hylobates* sp.).
- Notes: Within hylobatids, Kohlbrügge (1890–1892), Hepburn (1892), Jouffroy (1971), Gibbs (1999) and Michilsens et al. (2009) referred to an origin of the flexor digitorum superficialis from the ulna, radius and medial epicondyle of the humerus and/or from the common flexor tendon, as we found in our *H. lar* specimen HU HL1. A few researchers stated that there is no ulnar (e.g., Loth 1931) or radial (e.g., Deniker 1885) origin and there is neither a radial nor an ulnar origin in our *H. gabriellae* specimen VU HG1.

- Synonymy: Fléchisseur superficiel (Deniker 1885); flexor sublimis digitorum or flexor digitorum sublimis (Kohlbrügge 1890–1892, Hepburn 1892 and Grönroos 1903).

Palmaris longus (4.0 g; Figs. 27, 28, 31, 34)
- Usual attachments: Medial epicondyle of the humerus to the palmar aponeurosis.
- Usual innervation: Ulnar nerve only (Kohlbrügge 1890–1892; *H. moloch*, *H. agilis*, *H. syndactylus*); median nerve (Hepburn 1892: *Hylobates* sp.).
- Notes: In the hylobatids described by Kohlbrügge (1890–1892), Hepburn (1892), Grönroos (1903), Payne (2001), Michilsens et al. (2009) and Kikuchi (2010a,b), and dissected by us (2 specimens), the palmaris longus was always present and according to Loth (1931), Gibbs (1999) and Gibbs et al. (2002) it is present in 100% of hylobatids, although Deniker (1885) suggested that it was missing in the *Hylobates* sp. fetus dissected by him.
- Synonymy: Palmaire grêle (Deniker 1885); palmaris (Michilsens et al. 2009).

Flexor carpi ulnaris (3.8 g; Figs. 27, 31, 34)
- Usual attachments: From the ulna (caput ulnare) and sometimes (usually?) also from the medial epicondyle of the humerus (caput humerale) to the pisiform.
- Usual innervation: Median nerve (Kohlbrügge 1890–1892: *H. moloch*, *H. agilis*, *H. syndactylus*; Hepburn 1892: *Hylobates* sp.).
- Notes: Jouffroy (1971) stated that in hylobatids there is no direct bony origin from the ulna and according to Kohlbrügge (1890–1892) this was the case in the three hylobatids dissected by him (*H. moloch*, 1sp; *H. agilis*, 1 sp.; *H. syndactylus*, 1 sp.) and it was also the case in the 11 hylobatid specimens examined by Michilsens (2009; 3 *H. lar*, 2 *H. pileatus*, 2 *H. moloch* and 4 *H. syndactylus*). However, in the *Hylobates* sp. specimen reported by Hepburn (1892) and in the three hylobatids in which we examined this feature in detail (VU HG1, VU HG2 and HU HL1) the origin is from the ulna and the humerus and in her review Gibbs (1999) stated that an origin from both the humerus and the ulna is the usual condition for hylobatids. Therefore, until more information is available for hylobatids, it is not clear if the usual condition for these primates is to have an origin from the humerus or not. Kohlbrügge (1890–1892), Deniker (1885), Howell and Straus (1932) and Michilsens et al. (2009) did not found/report a distinct **epitrochleoanconeus** in hylobatids, and we did also not found this muscle in the hylobatids dissected by us.
- Synonymy: Cubital antérieur (Deniker 1885).

Flexor carpi radialis (8.4 g; Figs. 27, 31, 34)
- Usual attachments: From the medial epicondyle of the humerus and sometimes (usually?) also from the radius, to the base of metacarpal II and sometimes (usually?) also to the base of metacarpal III.

- Usual innervation: Median nerve (Kohlbrügge 1890–1892: *H. moloch, H. agilis, H. syndactylus*; Hepburn 1892: *Hylobates* sp.).
- Notes: According to Hepburn (1892) and Jouffroy (1971) in hylobatids the flexor carpi radialis goes to both metacarpals II and III and this is precisely the condition found in our *H. lar* specimen HU HL1. However, Michilsens et al. (2009) stated that, within the 11 hylobatids dissected by them (3 *H. lar*, 2 *H. pileatus*, 2 *H. moloch* and 4 *H. syndactylus*), the muscle always attaches to metacarpal II only and such an insertion was also found in the 3 hylobatids dissected by Kohlbrügge (1890–1892; 1 *H. agilis*, 1 *H. moloch* and 1 *H. syndactylus*) and in our *H. gabriellae* specimen VU HG1. Therefore, it is not clear if the usual condition for hylobatids is to have an insertion to metacarpal II only or to both metacarpals II and III. Regarding the origin of the flexor carpi radialis, according to Hepburn (1892; *Hylobates* sp., 1 specimen) and Kolhbrügge (1890–1892; 1 *H. agilis*, 1 *H. moloch* and 1 *H. syndactylus*) the muscle originates from both the humerus and the radius and this is the condition found in our *H. gabriellae* specimens VU HG1 and VU HG2 and in our *H. lar* specimen HU HL1. However, Michilsens et al. (2009) stated that in the 11 hylobatids dissected by them (3 *H. lar*, 2 *H. pileatus*, 2 *H. moloch* and 4 *H. syndactylus*) the muscle originated from the humerus only, except in the three specimens of *H. lar* in which the muscle also originated from the pronator teres and thus indirectly from the radius. Therefore, it is also not clear if the usual condition for hylobatids is to have an insertion onto the humerus only or instead onto both the humerus and radius.

Pronator teres (6.9 g; Figs. 27, 34)
- Usual attachments: From the humerus (caput humerale) and often (seemingly not usually: see notes below), also from the ulna, to the radius.
- Usual innervation: Median nerve (Kohlbrügge 1890–1892: *H. moloch, H. agilis, H. syndactylus*; Hepburn 1892: *Hylobates* sp.).
- Notes: The hylobatids reported by Hepburn (1892; *Hylobates* sp.), Chapman (1900; *H. moloch*), Lewis (1989; *H. lar*) and Michilsens et al. (2009; 3 *H. lar*, 2 *H. pileatus*, 2 *H. moloch* and 4 *H. syndactylus*) and our *H. gabriellae* specimens VU HG1 and VU HG2 only have a humeral origin of the pronator teres. Deniker (1885) suggested this was also the case in the *Hylobates* sp. fetus dissected by him, but the 3 specimens dissected by Kohlbrügge (1890–1892; 1 *H. agilis*, 1 *H. moloch* and 1 *H. syndactylus*), the specimen described by Stern and Larson (2001) and our *H. lar* specimen HU HL1 have a few fibers that also originate from the ulna (although there is no distinct well defined ulnar head such as that found in modern humans). According to Loth (1931) and Jouffroy (1971) a distinct ulnar head is only found in 41% and 42% of hylobatids, respectively. Ashton and Oxnard (1963), Ziegler (1964), and Stern and Larson (1993) commented on the particular importance of the pronator teres in brachiating,

where pronation of the forelimb, in suspended postures, causes this muscle to function as a primary flexor of the elbow.

- Synonymy: Rond pronateur (Deniker 1885); pronator radii teres (Hepburn 1892, Chapman 1900, Sonntag 1924).

Palmaris brevis (0.13 g; Figs. 28, 31)

- Usual attachments: From the pisiform and/or the flexor retinaculum to the skin of the medial border of the palm.
- Usual innervation: Data are not available.
- Notes: Hepburn (1892) stated that the *Hylobates* sp. specimen dissected by him had no palmaris brevis and this was also the case in the 3 specimens dissected by Kohlbrügge (1890–1892; 1 *H. agilis*, 1 *H. moloch* and 1 *H. syndactylus*); Loth (1931) stated that the muscle is always missing in hylobatids and Howell and Straus (1933) and Jouffroy (1971) wrote that it is usually absent or poorly developed in these primates. However, such statements are based on previous descriptions of other authors that are, in turn, based in very few dissections. In fact, in two of the three hylobatids dissected by us in which we examined this feature in detail (*H. lar* specimen HU HL1 and *H. gabriellae* specimen VU HG2) the muscle is present and well-developed and it was also present in the *H. leucogenys* illustrated by Dylevsky (1967), so according to our own review of the literature and of the data obtained in our own dissections, it is present in 3/8 of the cases (i.e., in about 38% of hylobatids). It is however not clear if the muscle was really missing in the other 5 specimens or whether it might have been removed with the skin, as is often the case with this little muscle in modern humans. Nevertheless, in view of the information available the usual condition for hylobatids is that in which the palmaris brevis is missing. The **flexor digitorum brevis manus** and **palmaris superficialis** (see, e.g., Diogo et al. 2009a) are usually not present as distinct muscles in hylobatids.
- Synonymy: Palmaire cutané (Deniker 1885).

Lumbricales (I [to digit 2]: 0.15 g; II [to digit 3]: 0.31 g; III [to digit 4]: 0.23 g; IV [to digit 5]: 0.07 g; Figs. 28, 29, 31, 34)

- Usual attachments: From the dorsal surfaces of the tendons of the flexor digitorum profundus to digits 2, 3, 4 and 5, to the radial side of the proximal phalanx and the extensor expansion of digits 2 (lumbricalis I), 3 (lumbricalis II), 4 (lumbricalis III) and 5 (lumbricalis IV, which is often, but not usually, missing in hylobatids).
- Usual innervation: First and second lumbricales by median nerve and third and fourth lumbricales by deep branch of ulnar nerve (Kohlbrügge 1890–1892: *H. moloch, H. agilis, H. syndactylus*; Hepburn 1892: *Hylobates* sp.).
- Notes: As explained by Tuttle (1969) in hylobatids the fourth lumbrical (i.e., the lumbrical going to digit 5) is frequently missing (it was missing in 7 of the *H. lar* hands, but was present in the 2 hands of *H. pileatus* and in all 9 hands

of *H. syndactylus* dissected by that author and it was present in the *Hylobates* sp. specimen described by Jouffroy and Lessertisseur (1960), the *Hylobates* sp. specimen reported by Hepburn (1892), the 3 hylobatids studied by Kohlbrügge (1890–1892; *H. moloch*, *H. agilis*, *H. syndactylus*) and in the *Hylobates* sp. fetus dissected by Deniker (1885) and it was also present in 2 of the four hylobatid hands in which we examined this feature in detail (*H. gabriellae* specimens VU HG1 and VU HG2). Thus, according to these numbers this muscle was missing in 10 out of 28 cases (i.e., in about 36% of the cases). Deniker (1885) stated that the fetal *Hylobates* sp. specimen dissected by him has 3 lumbricales to digits 3–5, but the structure that he designated as 'contrahens to digit 2' might well be a lumbricalis to digit 2 instead. As explained by Tuttle (1969) and corroborated by our dissections, in hylobatids the lumbricales mainly originate from the dorsal (and not from the ventral, or palmar) surfaces of the tendons of the flexor digitorum profundus. In our *H. lar* specimen HU HL1 the first, second and third lumbricales (the fourth is missing: see above) run from the tendons of the flexor digitorum to digits 2, 3 and 3+4, respectively, to the extensor expansion of digits 2, 3 and 4, respectively. On one side of the body of our *H. gabriellae* specimen VU GG1 the 1st lumbrical originates from the tendon of the flexor digitorum profundus to digit 2, the 2nd lumbrical from tendons to digits 2 and 3, the 3rd lumbrical from tendons to digits 3 and 4, and the 4th lumbrical 4 (which is very reduced in size, but present) from tendons of digits 4 and 5; on the other side of the body everything is similar, excepting that the 4th lumbrical is missing. The **intercapitulares** (see, e.g., Jouffroy 1971) are usually not present in hylobatids.

Contrahentes digitorum (I [to digit 2]: 0.37 g; II [to digit 4]: 0.18 g; III [to digit 5]: 0.14 g; Figs. 30, 32, 36)

- Usual attachments: Mainly from the contrahens fascia to the base of each proximal phalanx and to the extensor expansions of the ulnar side of digit 2 (contrahens I), of the radial side of digit 4 (contrahens II) and of the radial side of digit 5 (contrahens III).
- Usual innervation: Deep branch of ulnar nerve (Kohlbrügge 1890–1892: *H. moloch*, *H. agilis*, *H. syndactylus*; Hepburn 1892: *Hylobates* sp.); deep division of the ulnar nerve, which receives, in the forearm, a branch from the median nerve (Fitzwilliams 1910: *H. agilis*; this author states that these muscles are innervated by "the median nerve by way of the deep division of the ulnar nerve").
- Notes: Within hylobatids, in the specimens reported by Hartmann (1886; 1 *Hylobates* sp. specimen), Hepburn (1892; 1 *Hylobates* sp. specimen), Fitzwilliams (1910; 1 *H. agilis* specimen) and Lewis (1989; 1 *H. lar* specimen), one (the *H. moloch* specimen) of the three specimens dissected by Kohlbrügge (1890–1892), two (the 2 *H. moloch* specimens) of the three specimens dissected by Jouffroy

and Lessertisseur (1959) and the three hylobatids in which we examined this feature in detail (*H. gabriellae* specimens VU HG1 and VU HG2 and *H. lar* specimen HU HL1), there were contrahentes to digits 2, 4 and 5, as is seemingly also the case in the *Hylobates* sp. fetus dissected by Deniker (1885). This is the usual condition for hylobatids according to Jouffroy and Lessertisseur (1959) and Lewis (1989), although Day and Napier (1963) stated that in the *H. lar* specimen dissected by them there were no contrahentes other than the adductor pollicis, Bischoff (1870; 1 specimen of *H. moloch*) and Chapman (1900; 1 specimen of *H. moloch*) stated that in the specimens dissected by them there were contrahentes to digits 2 and 5 but not to digit 4 (as was also found in two—i.e., in the *H. syndactylus* and *H. agilis*—of the three specimens dissected by Kohlbrügge 1890–1892) and in one (i.e., the *H. lar* specimen) of the three specimens dissected by Jouffroy and Lessertisseur (1959) there was a single fleshy contrahens to digit 5. Therefore, according to our dissections and our review of the literature, in hylobatids there are contrahentes to digits 2, 4 and 5 in 10 out of 16 cases (i.e., in c.60% of the cases).

Adductor pollicis (1.0 g; Figs. 28, 29, 30, 32, 33, 36)

- Usual attachments: The caput obliquum and caput transversum are usually well-differentiated, connecting the metacarpals I, II and/or III, the contrahens fascia, and often at least some carpal bones and/or ligaments, to metacarpal I and to the metacarpophalangeal joint, as well as to the proximal phalanx and sometimes also to the distal phalanx of the thumb (directly by means of a thin tendon and/or indirectly by means of a ligament); when present, the TDAS-AD (see notes below) connects the proximo-medial portion of the metacarpal I and/or adjacent carpal structures to the metacarpophalangeal joint and/or the proximal portion of the proximal phalanx of digit 1 (see Fig. 29).

- Usual innervation: Deep branch of ulnar nerve (Kohlbrügge 1890–1892: *H. moloch*, *H. agilis*, *H. syndactylus*; Hepburn 1892: *Hylobates* sp.); deep division of the ulnar nerve, which receives, in the forearm, a branch from the median nerve (Fitzwilliams 1910: *H. agilis*; this author states that these muscles are innervated by "the median nerve by way of the deep division of the ulnar nerve").

- Notes: There has been much controversy regarding the homologies of the thenar muscles of primate and non-primate mammals. This subject has been discussed in detail in the recent studies of Diogo et al. (2009a) and particularly of Diogo and Abdala (2010) and here we summarize the main conclusions of those studies. The '**interosseous volaris primus of Henle**' of modern human anatomy corresponds very likely to a **thin, deep additional slip of the adductor pollicis** (**TDAS-AD** *sensu* Diogo and Abdala 2010), and not to the **flexor brevis profundus 2** of 'lower' mammals, as suggested by some authors (and by the erroneous use of the name 'interosseous volaris primus of Henle').

To our knowledge, the TDAS-AD has not been described in hylobatids, and is seemingly missing in the *H. gabriellae* specimen VU HG1 and in the *H. lar* specimen HU HL1. However, it is present in at least one hand of the *H. gabriellae* specimen VU HG2 (its muscle weight being 0.15 g; Fig. 29), and has also been found in other hylobatid specimen by other researchers (Sam Dunlap, pers. comm.). Regarding the '**deep head of the flexor pollicis brevis**' of modern human anatomy, this corresponds very likely to the flexor brevis profundus 2 of 'lower' mammals. We clearly found a flexor brevis profundus 2 in hylobatids, and this structure was seemingly also found in this taxon by Brooks (1887) and Kohlbrügge (1890–1892), although most authors designated this structure as 'deep head of the flexor pollicis brevis'. Regarding the '**superficial head of the flexor pollicis brevis**' of human anatomy, this thus seems to correspond to a true **flexor pollicis brevis**. This true flexor pollicis brevis and the **opponens pollicis** derive very likely from the **flexor brevis profundus 1** of 'lower' mammals, and are present as distinct muscles in the vast majority of hylobatids. Regarding the oblique and transverse heads of the adductor pollicis, some authors (e.g., Kohlbrügge 1890–1892; Sonntag 1924; Day and Napier 1963) suggest that these heads are often not differentiated in hylobatids, but both heads are clearly present in the hylobatids dissected by us and by authors such as Bischoff (1870), Fitzwilliams (1910), Deniker (1885), Hepburn (1892) and Duckworth (1904). In the specimens dissected by Brooks (1887), Kohlbrügge (1890–1892) and Duckworth (1904) the adductor pollicis is partially inserted onto the distal phalanx of the thumb. In hylobatids the adductor pollicis is usually partially inserted onto much of metacarpal I (i.e., functionally the muscle becomes an 'adductor' but also an 'opponens' of the thumb: e.g., Jouffroy and Lessertisseur 1960; our dissections). Tuttle (1969) stated that in hylobatids in association with the deep cleft (between digits 1 and 2) that nearly reaches the first intermetacarpal joint, the transverse head on the adductor pollicis muscle generally originates only in the proximal one-half or two-thirds of the palm, the majority of its distal fibers being directed parallel to the first metacarpal bone; thus the muscle may act as a flexor of the metacarpophalangeal joint when the thumb is adducted, while when the thumb is widely abducted it may act as an adductor/flexor of the carpometacarpal joint and flexor of the metacarpophalangeal joint (see his. Fig. 18). In our *H. lar* specimen HU HL1 the oblique head of the adductor pollicis seems to be less differentiated than in other hominoids, but it is clearly a distinct structure anyway. The adductor pollicis runs from the trapezoid, metacarpals II and III and flexor retinaculum to the ulnar side of the distal portion of metacarpal I (to a wide portion of this bone, but the fibers that attach there do not form a distinct structure, they are deeply blended with the remaining fibers of the muscle, so there is no distinct 'adductor opponens pollicis'), the metacarpophalangeal joint and the base of the proximal phalanx of digit 1; there is no tendon to the distal phalanx of this digit (the supposed tendon of this muscle going to this distal phalanx show in Fig. 5 of Brooks 1887 clearly seems to be the tendon of the flexor pollicis longus); the 'interosseous volaris primus' of modern human anatomy, i.e., the TDAS-AD *sensu* the present study, is not present as an independent structure. In our *H. gabriellae* specimen VU GG1 the oblique and transverse heads of the adductor pollicis were only slightly differentiated, the adductor pollicis running from metacarpals II and III, the flexor retinaculum, the trapezoid, and the contrahens fascia to the ulnar side of metacarpal I (but there is no distinct 'adductor opponens pollicis'), of the metacarpophalangeal joint and of the base of the proximal phalanx of digit 1, being only slightly connected to the ligament/tendon that passes from the proximal phalanx to the distal phalanx of digit 1. At least in this case this latter structure does seem to be a ligament, which passes medially (ulnar) to the tendon of the flexor digitorum profundus to digit 1 and there is no distinct TDAS-AD. Hepburn (1892) states that in the *Hylobates* sp. specimen dissected by him the adductor pollicis is associated with a fibrous aponeurosis and septum that runs distally from the base of metacarpal III, the transverse head inserting mainly onto the ulnar side of the base of the proximal phalanx of the pollex and onto the distal two thirds of the ulnar border of metacarpal I, and the oblique head originating from the palmar aspect of the base of metacarpal II and from the tendon of insertion of the flexor carpi radialis, in close proximity to the carpus. Brooks (1887) stated that in the *H. agilis* specimen dissected by him the adductor pollicis originates from metacarpal III and from the ligaments over the bases of metacarpals II and III and inserts onto the distal four-fifths of the ulnar side of the shaft of metacarpal I, the ulnar sesamoid bone of the thumb and the adjacent part of the proximal phalanx and also to the base of the distal phalanx by a slender aponeurotic slip. Kohlbrügge (1890–1892) stated that the adductor pollicis is mainly undivided, running from the contrahens fascia and metacarpals II and III to the distal 2/3 (*H. agilis* and *H. syndactylus*) or distal 1/3 (*H. moloch*) of metacarpal I, the sesamoid bone, the base of the proximal phalanx of the thumb and, through a thin slip, also to the distal phalanx of the thumb. Duckworth (1904) reported one *H. muelleri* specimen in which the transverse head of the adductor pollicis originates from the proximal half of the shaft of metacarpal III and he stated that in hylobatids this head is often inserted onto the distal and proximal phalanges of the thumb, the oblique head is usually also inserted onto the distal phalanx, and there is, frequently, a slip from the adductor pollicis to metacarpal I, which is described as an 'adductor opponens pollicis'; according to him the occurrence of a slip from the oblique head of the adductor pollicis to the radial sesamoid bone of the thumb is infrequent in hylobatids. Fitzwilliams (1910) reported one *H. agilis* specimen in which the transverse head of the adductor pollicis runs from the flexor retinaculum and the bases of metacarpals II and III to the ulnar side of the distal half of metacarpal I, while the oblique head of this muscle runs from the proximal half of metacarpal III and the fascia over the metacarpals to the ulnar side of the distal half of metacarpal I and the ulnar side of the proximal phalanx of the thumb.

- Synonymy: Contrahens I (Fitzwilliams 1910).

Interossei palmares (I [to digit 2]: 0.56 g; II [to digit 4]: 0.35 g; III [to digit 5]: 0.23 g; Figs. 30, 32, 33, 35)

- Usual attachments: Interosseous palmaris I runs mainly from metacarpal II to the ulnar side of the proximal phalanx and extensor expansion of digit 2; interosseous palmaris II mainly from metacarpal IV to the radial side of the proximal phalanx and extensor expansion of digit 4; interosseous palmaris III mainly from metacarpal V to the radial side of the proximal phalanx and extensor expansion of digit 5.

- Usual innervation: In the *H. agilis* specimen dissected by Brooks (1887), the *Hylobates* sp. specimen dissected by Hepburn (1892) and the *H. moloch, H. agilis,* and *H. syndactylus* specimens dissected by Kohlbrügge (1890–1892) the interossei palmares I, II and III are innervated by the deep branch of the ulnar nerve; Fitzwilliams (1910) stated that in *H. agilis* the interossei palmares I, II and III are innervated by the deep division of the ulnar nerve, which receives, in the forearm, a branch from the median nerve (Fitzwilliams stated that these muscles are innervated by "the median nerve by way of the deep division of the ulnar nerve").

- Notes: In the hylobatids dissected by us the four dorsal interossei and the second, third and fourth in particular, have two distinct bundles, one passing mainly superficially to the transverse lamina of the metacarpophalangeal joint (this bundle is the one that gives rise to the interossei accessorii) and the other passing mainly deep to this lamina. The bundle passing superficially to the lamina thus clearly corresponds to a flexor brevis profundus of the chimpanzee, while the one passing deep to the lamina corresponds to an intermetacarpalis of the chimpanzee. Within the first dorsal interosseous of hylobatids one can distinguish these two bundles distally, because one of them is superficial and the other deep to the transverse lamina, but proximally the two bundles are deeply blended, clearly forming a dorsal interosseous similar to that of modern humans. That is why we consider that hylobatids have dorsal and palmar interossei, as modern humans do, and not intermetacarpales and flexores breves profundi such as those found in chimpanzees (in the chimpanzees, even the intermetacarpalis and the flexor brevis profundus that go to the radial side of digit 2—i.e., that form, in other hominoids, the second dorsal interosseous —are well separated from each other). Thus, one can say that hylobatids have 3 palmar interosseous and 4 dorsal interosseous, but that the intermetacarpales and flexores breves profundi that form the dorsal interossei, particularly the dorsal interossei 2, 3 and 4, are not as fused to each other as they are in modern humans. The dorsal interossei 1, 2, 3 and 4 of our hylobatid specimens run mainly from metacarpals II, II+III, III+IV and IV+V, respectively, to the extensor expansion and the region near the base of the proximal phalanx of the radial side of digit 2, of the radial side of digit 3, of the ulnar side of digit 3, and of the ulnar side of digit 4, respectively (see also interossei accessorii below); the palmar interossei 1, 2, and 3 of hylobatids run mainly from metacarpals II, IV and V, respectively, to the extensor expansion and the region near the base of the proximal phalanx of the ulnar side of digit 2, of the radial side of digit 4, and of the radial side of digit 5, respectively. Kohlbrügge (1890– 1892) stated that in *H. moloch*, *H. agilis* and *H. syndactylus* there are 3 'interossei interni' to the ulnar side of digit 2 and to the radial sides of digits 4 and 5 (which correspond to the interossei palmares to digits 2, 4 and 5 *sensu* the present study, i.e., to the flexores breves profundi 4, 7 and 9 of other mammals); in addition he described a 'caput ulnare of the flexor pollicis brevis' running from the flexor retinaculum to the sesamoid bone, but also to the distal 3/4 of metacarpal I in *H. agilis* (which probably corresponds to the 'deep head of the flexor pollicis brevis' of modern human anatomy and thus to the flexor brevis profundus 2 *sensu* the present study). According to him the flexores breves profundi 3, 5, 6 and 8 are thus fused with the intermetacarpales to form the four dorsal interossei ('interossei externi' *sensu* Kohlbrügge), which go to the radial sides of digits 2 and 3 and the ulnar sides of digits 3 and 4; however, as stressed by him these dorsal interossei can be easily divided into a 'palmar' bundle and a 'dorsal' bundle corresponding respectively to the portion of the flexores breves profundi and to the intermetacarpales that form these muscles (see Fig. 5 of his Plate 18; N.B. Kohlbrügge 1890–1892 also described an 'interosseous accessorius' that extends to the radial side of the distal phalanx of digit 5, which he designated as 'abductor tertii internodii indicis': see Fig. 4 of his Plate 18). Fitzwilliams (1910) stated that in the *H. agilis* specimen dissected by him there are two palmar interossei, one running from metacarpal II and slightly from the base of metacarpal III to the ulnar side of the base of the proximal phalanx of digit 2 and the dorsal extensor expansion of that digit, the other running from metacarpal V to the radial side of the base of the proximal phalanx of digit 5 and the ligaments of the metacarpophalangeal joint of that digit. He stated that there was no palmar interosseous to digit 4 and described four interossei accessorii, which, according to him, correspond to four flexores breves profundi (i.e., in this case the four 'dorsal interossei' would not be true dorsal interossei, but instead four intermetacarpales): the first runs from metacarpal II, between the palmar and dorsal interossei of these digits and closely associated with them, to the radial side of the middle phalanx of digit 2, with some tendinous processes extending distally to the pulp of this digit; the second, which is partly segmented into two bellies, runs from metacarpal III (the posterior and anterior bellies) and possibly metacarpal II (the posterior belly) to the radial side of the proximal phalanx (the posterior belly) and the extensor expansion (the posterior and anterior bellies) of digit 3; the third, which is also partly segmented into two bellies, runs from metacarpal III (the posterior and anterior bellies) and the fascia that gives origin to the contrahentes (the anterior belly) to the ulnar side of the proximal phalanx (the posterior belly) and the extensor expansion (the anterior belly) of digit 3; the fourth runs from metacarpal IV to the ulnar side of the proximal phalanx and the extensor expansion of digit 4 (thus, these four muscles interossei accessorii and the four dorsal interossei occupy the same position relative to the digits (radial side of digits 2 and 3 and ulnar side of digits 3 and 4). Tuttle (1969) stated that in hylobatids, in addition to the usual insertion onto the ventrolateral base of the proximal phalanx of digit 2, the first dorsal interosseous muscle has a prominent accessory belly ('musculus accessorius interosseous 1') that extends distally to insert onto the base, and sometimes onto the ventrolateral aspect of the proximal shaft of the middle phalanx of the digit. Thus according to him the proximal portion of the first dorsal interosseous serves as a flexor of the proximal phalanx, while the 'musculus accessorius' may act as a flexor of the middle phalanx; the first dorsal interosseous generally has no insertion into the lateral band of the extensor sheath, but a fine tendon sometimes may be observed extending from the main insertion of the 'musculus accessorius interosseous 1' to the pulp of the finger. Also according to him, the bulk of the second, third, and fourth dorsal interossei insert onto the base of the proximal phalanx, each muscle also sending a fleshy extension distally as the 'musculus accessorius interosseous'; these insert onto the extensor sheath and into the proximal one third of the shaft of the proximal phalanx, the minor fasciculi of the dorsal interossei proper also contributing, occasionally, to the extensor sheaths (see his Fig. 16). Fig. 9.6A of Lewis (1989) shows a *Hylobates lar* specimen that seems to have all the 10 flexores breves profundi, including the second one. Susman et al.'s (1982) Fig. 1 shows a *H. syndactylus* specimen with a 'dorsal interosseous', a 'flexor brevis' and an 'accessorius interosseous' going to the radial side of digit 2, thus suggesting that in *Hylobates* all, or almost all, the flexores breves profundi are present as independent structures, and are not fused with the intermetacarpales. Susman et al. (1982) stated that in the two *H. syndactylus* and the two *H. lar* specimens dissected by them a 'true distally migrated and hypertrophied muscle accessorius interosseous' was found only on the radial side of digit 2, running from the palmar surface of metacarpal II to the extensor expansion at, or just distal to, the proximal interphalangeal joint. Contrary to Fitzwilliams (1910) and Huxley (1871), who stated that the muscle inserts onto the pulp of digit 2, and to Kohlbrügge (1890–1892), who stated that this muscle reaches the distal phalanx of digit 2 in gibbons and the middle phalanx in siamangs, in the gibbons and the siamangs dissected by

Susman et al. (1982) the tendon of insertion does not reach the distal phalanx of digit 2. Susman et al.'s (1982) Fig. 1 strongly suggests that this latter muscle is derived from the third flexor brevis profundus (i.e., the one going to the radial side of digit 2) and not from the intermetacarpales and these authors suggest that this interpretation is given support by the findings of Kohlbrügge (1890–1892), Fitzwilliams (1910) and Forster (1917). See also notes about adductor pollicis and about interossei dorsales.

Interossei dorsales (I [to digit 2]: 1.09 g; II [to digit 3, radial side]: 0.99 g; III [to digit 3, ulnar side]: 0.92 g; IV [to digit 4]: 0.69 g; Figs. 30, 32, 33, 35, 36)

- Usual attachments: Interosseous dorsalis I runs mainly from metacarpal II and sometimes also from metacarpal I, to the radial side of the proximal phalanx and extensor expansion of digit 2; interosseous dorsalis II mainly from metacarpals II and III to the radial side of the proximal phalanx and extensor expansion of digit 3; interosseous dorsalis III mainly from metacarpals III and IV to the ulnar side of the proximal phalanx and extensor expansion of digit 3; interosseous dorsalis IV mainly from metacarpals IV and V to the ulnar side of the proximal phalanx and extensor expansion of digit 4.
- Usual innervation: Deep branch of ulnar nerve (Kohlbrügge 1890–1892: *H. moloch, H. agilis, H. syndactylus*; Hepburn 1892: *Hylobates* sp.); deep division of the ulnar nerve, which receives, in the forearm, a branch from the median nerve (Fitzwilliams 1910: *H. agilis*; this author states that these muscles are innervated by "the median nerve by way of the deep division of the ulnar nerve", but that, contrary to the other dorsal interossei, the fibers that supply the fourth dorsal interosseous travel all the way in the ulnar nerve, i.e., they do not come from the median nerve).
- Notes: As explained above (see adductor pollicis and interossei palmares), the **flexor brevis profundus 2** of 'lower' mammals corresponds very likely to the '**deep head of the flexor pollicis brevis**' of modern human anatomy; the '**superficial head of the flexor pollicis brevis**' of modern human anatomy, as well as the **opponens pollicis**, derive very likely from the **flexor brevis profundus 1** of 'lower' mammals, while the **flexor digiti minimi brevis** and the **opponens digiti minimi** derive very likely from the **flexor brevis profundus 10** of 'lower' mammals (for a recent review, see Diogo et al. 2009a and Diogo and Abdala 2010). In hylobatids and other hominoids except *Pan* the **flexores breves profundi** 3, 5, 6 and 8 are fused with the **intermetacarpales** 1, 2, 3 and 4, forming the interossei dorsales 1, 2, 3 and 4, respectively; the **interossei palmares** 1, 2 and 3 of hylobatids, modern humans, gorillas and orangutans thus correspond respectively to the flexores breves profundi 4, 7 and 9 of 'lower' mammals (see, e.g., Lewis 1989, Diogo et al. 2009a, and Diogo and Abdala 2010). However, in at least some of the interossei dorsales of the hylobatids dissected by us it is still possible to identify the portions derived from the intermetacarpales and the the flexores breves

profundi (see notes about interossei palmares above). As noted by authors such as Brooks (1887) and Susman et al. (1982) and corroborated by our dissections, in hylobatids the interosseous dorsalis 1 usually originates from metacarpal II only (and not from both metacarpals I and II as is often the case in other hominoids; see interossei palmares above for the description of the interossei dorsales of the hylobatids dissected by us). However, some authors also found an interosseous dorsalis I origin from metacarpal I. For instance, Fitzwilliams (1910) stated that in the *H. agilis* specimen dissected by him there are four interossei dorsales: the first running from metacarpals I and II to the radial side of the proximal phalanx and partially onto the dorsal extensor expansion of digit 2; the second runs from metacarpals II and III to the radial side of the proximal phalanx and to the extensor expansion of digit 3; the third runs from metacarpals III and IV to the ulnar side of the proximal phalanx and to the extensor expansion of digit 3; the fourth runs from metacarpals IV and V to the ulnar side of the proximal phalanx and to the extensor expansion of digit 4. The **interdigitales** present in primates such as lorisoids, are usually not present in hylobatids.

- Synonymy: Interossei externi (Kohlbrügge 1890–1892); the first dorsal interosseous *sensu* the present study corresponds to the abductor indicis *sensu* Brooks (1887) and Fitzwilliams (1910).

Interossei accessorii (I [to digit 2]: 1.0 g; Figs. 28, 29, 30, 32, 36)
- Usual attachments: See notes below.
- Usual innervation: Deep branch of ulnar nerve (Kohlbrügge 1890–1892: *H. moloch*, *H. agilis*, *H. syndactylus*; Hepburn 1892: *Hylobates* sp.); deep division of the ulnar nerve, which receives, in the forearm, a branch from the median nerve (Fitzwilliams 1910: *H. agilis*; this author states that these muscles are innervated by "the median nerve by way of the deep division of the ulnar nerve").
- Notes: The structures that are designated as interossei accessorii in hylobatids are seemingly mainly derived from the distal portion of the flexores breves profundi/intermetacarpales; in these primates only the interosseous accessorius to digit 2 is usually well-developed. Gibbs (1999), based on Brooks (1887), Kohlbrügge (1890–1892), Fitzwilliams (1910), Forster (1917) and Susman et al. (1982) stated that hylobatids have an accessory interosseous muscle originating from the first dorsal interosseous or from the flexor pollicis brevis and inserting onto the base and ventrolateral shaft of the middle phalanx of digit 2, sometimes with a small fleshy extension to the pulp of the digit, according to Fitzwilliams (1910). The muscle may insert onto the extensor expansion at or just distal to the proximal interphalangeal joint according to Fitzwilliams (1910) and Susman et al. (1982) and appears to have a more extensive insertion in gibbons than in siamangs, reaching the distal phalanx in gibbons according to Brooks

(1887) and Kohlbrügge (1890–1892), but only the distal end of the middle phalanx in siamangs according to Kohlbrügge (1890–1892). Susman et al. (1982) showed that, during locomotion, the largest EMG potentials recorded in the interosseous accessorius to digit 2 of the two *H. lar* specimens analyzed by them was during brachiation along the overhead ladder: in both resisted and unresisted grasping of food, the activity of this muscle exceeded that observed in locomotor activities. According to them, the muscle is mainly an abductor of digit 2 during the pinch grasp of small objects while it also helps to flex the metacarpophalangeal joint during whole hand grasping. Also according to Susman et al. (1982), the wide separation between digits 1 and 2 in hylobatids lead to evolutionary advantages; it also resulted in a reduction in size of the 'first dorsal interosseous' muscle (and of the adductor pollicis that contrary to the condition in other hominoids in hylobatids provides resistance to the thumb in pinch grasping; in hylobatids the first dorsal interosseous usually does not originate from metacarpal I). The interosseous accessorius thus assumes much of the role of abduction that in other hominoids would normally fall to the part of the 'first dorsal interosseous' that arises from metacarpal I. In the hylobatids dissected by us the interossei accessorii seem to be extensions of the part of the dorsal interossei that passes superficially to the transverse lamina of the metacarpophalangeal joint (i.e., they seem to be extensions of the part that corresponds to the flexores breves profundi; the interossei accessorii associated with the dorsal interossei 2, 3 and 4 do not extend much further than the base of the proximal phalanges of digits 3, 3 and 4, respectively, but that associated with the dorsal interosseous 1 does extend further distally to attach onto the middle phalanx (but not onto the distal phalanx) of digit 2. That is, there is only one well-developed 'musculus accessorius interosseous 1' (going to digit 2) as defended by authors such as Susman et al. (1982).

- Synonymy: Abductor tertii internodii secundi digiti or abductor tertii internodii indicis (Huxley 1871, Kohlbrügge 1890–1892); part of interossei (Hartmann 1886); extensor tertii internodi indicis (Keith 1894a); interosseous volaris radialis longus (Forster 1917, 1933).

Flexor pollicis brevis and *flexor brevis profundus 2* ('deep head' plus 'superficial head': 0.8 g; Figs. 28, 32, 33, 36)

- Usual attachments: As explained above (see adductor pollicis), the flexor brevis profundus 2 of 'lower' mammals corresponds very likely to the **deep head of the flexor pollicis brevis**' of modern human anatomy; the **superficial head of the flexor pollicis brevis**' of modern human anatomy thus corresponds to a true flexor pollicis brevis, being very likely derived, together with the **opponens pollicis**, from the **flexor brevis profundus 1** of 'lower' mammals (for a recent review, see Diogo et al. 2009a and Diogo and Abadala 2010). As there is much confusion in the literature about the so-called 'superficial and deep heads of the flexor pollicis brevis' in hylobatids (and other primates), we will describe here in some detail the condition found in our *H. lar* specimen HU HL1 and in our *H. gabriellae* specimen VU HG1. As shown in Fig. 16 of Jouffroy and Lessertisseur (1960) in our HU HL1 specimen the structure that is often named 'flexor pollicis brevis' clearly has two 'heads' that are well-separated distally but somewhat blended proximally and which clearly seem to correspond to the 'superficial and deep heads of the flexor pollicis brevis' of modern human anatomy (i.e., to part of the flexor brevis profundus 1 and to the flexor brevis profundus 2 *sensu* the present study, respectively). Proximally these two 'heads' originate mostly from the flexor retinaculum and the trapezium; distally the 'superficial head' inserts onto the radial side of the metacarpophalangeal joint and the base of the proximal phalanx of digit 1, together with the abductor pollicis brevis, while the 'deep head' inserts mostly on the ulnar side of the metacarpophalangeal joint and the base of the proximal phalanx of this digit. The so-called 'flexor pollicis brevis', and particularly its 'deep head', apparently sends some slips (but not a well-defined tendon) to the distal phalanx of digit 1. On both sides of our specimen VU HG1 there is a 'deep head of the flexor pollicis brevis' of modern human anatomy' (i.e., flexor brevis profundus 2 *sensu* the present study) running mainly from the flexor retinaculum to the ulnar side of the metacarpophalangeal joint and base of proximal phalanx of digit 1, passing dorsally to the tendon of the flexor digitorum profundus to digit 1; the 'superficial head of the flexor pollicis brevis of human anatomy' (flexor pollicis brevis *sensu* the present study) originates mainly from the flexor retinaculum and inserts onto the radial side of the metacarpophalangeal joint and to the base of proximal phalanx of digit 1 (there are seemingly no fibers/tendons of any of these two structures to the distal phalanx of the thumb). In the *Hylobates* sp. specimen dissected by Hepburn (1892) the 'deep head of the flexor pollicis brevis' of modern human anatomy, i.e., the flexor brevis profundus 2 *sensu* the present study (note that in 1892 he designated this structure as the 'interosseous volaris primus of Henle', but then in 1896 he clearly stated that this structure is not homologous to the 'interosseous volaris primus of Henle' of modern human anatomy and defended, as we do in the present study, that the flexor brevis profundus 2 corresponds to the 'deep head of the flexor pollicis brevis' of modern human anatomy) arises deeply in the palm from the ligamentous structures in the vicinity of the trapezium and from the bases of metacarpals I and II and inserts in conjunction with the adductor pollicis. Deniker (1885) stated that in the fetal *Hylobates* sp. specimen dissected by him there are two 'heads of the flexor pollicis brevis' so this fetus has a flexor brevis profundus 2 *sensu* the present study. Brooks (1887) reported one *H. agilis* specimen in which the 'superficial head of the flexor pollicis brevis' runs from the flexor retinaculum and metacarpal I to the sesamoid bone of the radial side of the metacarpophalangeal joint of digit 1, sending a small slip to the distal phalanx of this digit; the 'interosseous volaris primus of Henle' (which in this case corresponds to the 'deep head of the flexor pollicis brevis' of modern human anatomy and thus to the flexor brevis profundus 2 *sensu* the present study) runs from the metacarpal I and the flexor retinaculum to the sesamoid bone of the ulnar side of the metacarpophalangeal joint of digit 1. Kohlbrügge (1890–1892) dissected *H. moloch*, *H. agilis* and *H. syndactylus* specimens and stated that the 'caput radiale of the flexor pollicis brevis' (which corresponds to the 'superficial head of the flexor pollicis brevis' of modern human anatomy) runs from the flexor retinaculum to the distal portion of metacarpal I (only in *H. agilis* and *H. syndactylus*), to the sesamoid bone and the center of the base of the proximal phalanx of the thumb together with the abductor pollicis brevis (in the three species) and also through a small slip to the distal phalanx of the thumb (only in *H. agilis* and *H. moloch*). Fitzwilliams (1910) reported one *H. agilis* specimen in which the 'flexor pollicis brevis' has a 'superficial head' originating from the scaphoid, trapezium, sesamoid bone and the flexor retinaculum, and a 'deep head' originating from the trapezium and metacarpal I and corresponding to the flexor brevis profundus 2 *sensu* the present study; the 'flexor pollicis brevis' inserts onto the sesamoid bone of the metacarpophalangeal joint of digit 1 as well as onto the radial side of the shaft of the proximal phalanx of this digit, while the 'superficial head' is partly fused with the opponens pollicis. Tuttle (1969) stated that in hylobatids the 'flexor pollicis brevis' has a fleshy insertion to the proximal shaft of the proximal phalanx of digit 1, but occasionally a weak tendon continues distally from the phalangeal insertion to the pulp over the distal phalanx of digit 1.

- Usual innervation: In the *H. agilis* specimen dissected by Brooks (1887) and the *Hylobates* sp. specimen dissected by Hepburn (1892) the structure designated as 'interosseous volaris primus of Henle' and that corresponds to the 'deep head of the flexor pollicis brevis' of modern human anatomy and thus to the flexor brevis profundus 2 *sensu* the present study, is innervated by the median nerve; Kohlbrügge (1890–1892) stated that in *H. moloch*, *H. agilis*, and *H. syndactylus* the 'caput ulnare of the flexor pollicis brevis', which probably corresponds to the flexor brevis profundus 2 *sensu* the present study, is innervated by the median nerve. According to Brooks (1887: *H. agilis*), Kohlbrügge (1890–1892: *H. moloch*, *H. agilis*, *H. syndactylus*), and Hepburn (1892: *Hylobates* sp.) the 'superficial head of the flexor pollicis brevis' of modern human anatomy (i.e., the flexor pollicis brevis *sensu* the present study) is innervated by the median nerve.
- Notes: See notes about adductor pollicis, interossei palmares and interossei dorsales.

Opponens pollicis (0.6 g; Figs. 28, 32, 33, 35)
- Usual attachments: From the flexor retinaculum, trapezium and often the adjacent sesamoid bone and/or the scaphoid to the whole length of metacarpal I and often also to the proximal phalanx of the thumb.

- Usual innervation: Median nerve (Kohlbrügge 1890–1892: *H. moloch, H. agilis, H. syndactylus*; Hepburn 1892: *Hylobates* sp.); lateral division of the median nerve (Fitzwilliams 1910: *H. agilis*).
- Notes: Within hylobatids the insertion of the opponens pollicis extends to the proximal phalanx of the thumb in the *H. agilis* specimen described by Fitzwilliams (1910) and in the *H. syndactylus* specimen reported by Kohlbrügge (1890–1892) and to both the proximal and distal phalanges of the thumb in the *Hylobates* sp. specimen described by Hepburn (1892), but it extends only to the distal portion of the metacarpal I in the *H. agilis* specimens described by Brooks (1887) in the *H. agilis* specimen and the *H. moloch* specimen dissected by Kohlbrügge (1890–1892) and in the hylobatids dissected in this study.

Flexor digiti minimi brevis (0.5 g; Figs. 28, 32, 36)

- Usual attachments: From the hamate and/or the flexor retinaculum and sometimes also from pisiform, to the ulnar side of proximal phalanx and often also to the middle and distal phalanges of digit 5.
- Usual innervation: Deep branch of ulnar nerve (Kohlbrügge 1890–1892: *H. moloch, H. agilis, H. syndactylus*; Hepburn 1892: *Hylobates* sp.; Fitzwilliams 1910: *H. agilis*).
- Notes: Within hylobatids, an origin of the flexor digiti minimi brevis from the pisiform, flexor retinaculum and hamate was found by Fitzwilliams (1910; *H. agilis*) and by us in our *H. lar* specimen (but not in our *H. gabriellae* specimen VU HG2) and Kohlbrügge (1890–1892) suggested that in one of the three specimens dissected by him there is also an origin from the pisiform shared with the abductor digiti minimi. Hepburn (1892) described an origin from the flexor retinaculum and hamate in only one *Hylobates* sp. specimen and Deniker (1885) referred only to an origin from the hamate in the *Hylobates* sp. fetus dissected by him. The insertion of the muscle was reported as extending to the proximal and middle phalanges of digit 5 in the *H. agilis* specimen described by Fitzwilliams (1910) and in the specimens reported by Tuttle (1969) and to the distal porton of the proximal phalanx of this digit in the *Hylobates* sp. specimens reported by Hepburn (1892) and Sonntag (1924). Kohlbrügge (1890–1892) suggested that this was also the case in at least some of the hylobatids dissected by him although Deniker (1885) only referred to an insertion onto the base of the proximal phalanx in the *Hylobates* sp. fetus dissected by him. In the hylobatid specimens dissected by us it was not possible to discern if the insertion also extended distally to the base of the proximal phalanx of digit 1.
- Synonymy: Flexor brevis digiti quinti (Kohlbrügge 1890–1892).

Opponens digiti minimi (0.19 g; Figs. 30, 32, 35)

- Usual attachments: From the hamate and, often, also from the flexor retinaculum to the whole length of metacarpal V; occasionally the muscle may also originate from a portion of metacarpal V.

- Usual innervation: Deep branch of ulnar nerve (Kohlbrügge 1890–1892: *H. moloch, H. agilis, H. syndactylus*; Hepburn 1892: *Hylobates* sp.; Fitzwilliams 1910: *H. agilis*).
- Notes: Hepburn (1892) stated that in the *Hylobates* sp. specimen dissected by him the opponens digiti minimi runs from the hook of the hamate and from the flexor retinaculum to the ulnar border of the shaft of metacarpal V and is mainly superficial to the deep ulnar nerve and vessels. Deniker (1885) stated that in the *Hylobates* sp. fetus dissected by him the opponens digiti minimi runs from the hamate to metacarpal V. Kohlbrügge (1890–1892) stated that the opponens digiti minimi is mainly undivided in *H. syndactylus*, running from the flexor retinaculum and the base of metacarpal V to a more distal portion of that metacarpal, while in *H. agilis* and *H. moloch* the muscle is divided into an 'upper', well-developed bundle running from the hamate to the whole metacarpal V and a poorly-differentiated 'lower' bundle running from the hamate and the base of metacarpal V to a more distal portion of this metacarpal, the two bundles being only slightly divided by the deep branch of the ulnar nerve. Fitzwilliams (1910) reported one *H. agilis* specimen in which the opponens digiti minimi runs from the hook of the hamate, the flexor retinaculum and the anterior carpal ligaments to the ulnar side of metacarpal V, the muscle being partly blended with the palmar interosseous and the contrahens to digit 5. Lewis (1989) stated that the opponens digiti minimi is more markedly divided in hominoids such as *Pan* and modern humans than in hominoids such as hylobatids, and Brooks (1886a) stated that, contrary to *Pan* and modern humans, in hominoids such as *Pongo* there are no superficial and deep bundles of the muscle separated by the deep branch of the ulnar nerve. With respect to our dissections, in hylobatids, *Gorilla* and *Pongo* the deep branch of the ulnar nerve runs mainly radial to both these bundles and not mainly superficially (palmar) to the deep bundle and deep (dorsal) to the superficial bundle as is usually the case in *Pan* and particularly in modern humans. In our *H. lar* specimen HU HL1 and our *H. gabriellae* specimen VU HG1 the opponens digiti minimi runs from the hook of the hamate and the flexor retinaculum to the whole length of the ulnar side of metacarpal V; the muscle seems to be even more proximodistally oriented than in other non-human hominoids, having a distinct tendon that runs proximodistally to attach to the ulnar side of the distal margin of metacarpal V and that is in fact very similar to the tendon of the flexor digiti minimi brevis, thus providing further evidence that these two muscles are phylogenetically derived from the flexor brevis profundus 10 (see, e.g., Diogo et al. 2009a).
- Synonymy: Opponens digiti quinti (Kohlbrügge 1890–1892); opponens minimi digiti (Lewis 1989).

Abductor pollicis brevis (0.7 g; Figs. 28, 32, 33, 36)
- Usual attachments: From the flexor retinaculum and sometimes from the trapezium, trapezius, the adjacent sesamoid bone and/or the scaphoid, to the

radial side of the metacarpophalangeal joint and its sesamoid bone and to the base of the proximal phalanx of digit 1 and occasionally also to the distal phalanx of this digit and/or to the distal portion of metacarpal I.

- Usual innervation: Median nerve (Brooks 1887: *H. agilis*; Kohlbrügge 1890–1892: *H. moloch, H. agilis, H. syndactylus*; Hepburn 1892: *Hylobates* sp.; Fitzwilliams 1910: *H. agilis*).
- Notes: Hepburn (1892) stated that in the *Hylobates* sp. specimen dissected by him the abductor pollicis brevis runs from the flexor retinaculum, scaphoid and the sesamoid bone ('prepollex') to the radial side of the base of the proximal phalanx of the pollex and to the head of metacarpal I. Kohlbrügge (1890–1892) stated that in *H. moloch, H. agilis* and *H. syndactylus* the abductor pollicis brevis runs from the carpal region, flexor retinaculum and surrounding ligaments and also from the 'prepollex' in *H. agilis* to the proximal phalanx and sesamoid bone of the thumb and in *H. agilis* also to the metacarpal I. Fitzwilliams (1910) described an *H. agilis* specimen in which the abductor pollicis brevis runs from the sesamoid bone (or 'prepollex') and the ligaments between that bone, the scaphoid and the trapezium, as well as from the flexor retinaculum and the common head of origin of the opponens pollicis and the 'superficial head of the flexor pollicis brevis' of human anatomy to the radial side of the base of the proximal phalanx of the thumb. Tuttle (1969) stated that in gibbons the insertion of the well-developed abductor pollicis brevis is generally limited to the ventrolateral base of the proximal phalanx of digit 1 and that an extensor sheath is usually not evident in this digit. In our *H. lar* specimen HU HL1 and our *H. gabriellae* specimen VU HG1 the undivided abductor pollicis brevis runs from ligaments connecting the trapezium to the associated sesamoid bone and to the scaphoid, as well as from that sesamoid bone, the trapezius and the flexor retinaculum, to the radial side of the base of the proximal phalanx of digit 1; distally the muscle receives some fibers of the 'superficial head of the flexor pollicis brevis' of modern human anatomy (which corresponds to the flexor pollicis brevis *sensu* the present study).

Abductor digiti minimi (0.9 g; Figs. 28, 32, 36)
- Usual attachments: From the pisiform and sometimes from the flexor retinaculum and/or hamate, to the ulnar side of proximal phalanx of digit 5 and sometimes to the adjacent metacarpophalangeal joint and/or the base of the distal portion of metacarpal V.
- Usual innervation: Deep branch of ulnar nerve (Kohlbrügge 1890–1892: *H. moloch, H. agilis, H. syndactylus*; Hepburn 1892: *Hylobates* sp.; Fitzwilliams 1910: *H. agilis*).
- Notes: Hepburn (1892) described one *Hylobates* sp. specimen in which the abductor digiti minimi runs from the pisiform bone to the ulnar aspect of the base of the proximal phalanx of digit 5, being closely blended with the

insertion of the flexor digiti minimi brevis. Deniker (1885) reported a *Hylobates* sp. fetus in which the abductor digiti minimi inserts onto the distal portion of metacarpal V. Kohlbrügge (1890–1892) stated that in *H. moloch, H. agilis,* and *H. syndactylus* the abductor digiti minimi runs from the pisiform to the ulnar side of the proximal phalanx of digit 1, the muscle being only slightly differentiated into two bundles in *H. agilis* and *H. syndactylus*. Fitzwilliams (1910) stated that in the *H. agilis* specimen dissected by him the abductor digiti minimi runs from pisiform and the flexor retinaculum (one head of origin) and the ulnar side of the hamate (the other head of origin) to the ulnar side of the base of the proximal phalanx of digit 5. In our *H. lar* specimen HU HL1 and our *H. gabriellae* specimen VU HG1 the abductor digiti minimi runs from the pisiform and flexor retinaculum to the ulnar side of the base of the proximal phalanx of digit 5, being blended, distally, with the flexor digiti minimi brevis.

- Synonymy: Abductor digiti quinti (Kohlbrügge 1890–1892); abductor minimi digiti (Lewis 1989).

Extensor carpi radialis longus (3.6 g; Figs. 27, 34)

- Usual attachments: From the lateral supracondylar ridge and sometimes also from lateral epicondyle/supracondylar ridge of the humerus, to the base of metacarpal II and often (seemingly usually), also to base of metacarpal I.
- Usual innervation: Radial nerve (Kohlbrügge 1890–1892: *H. moloch, H. agilis, H. syndactylus*; Hepburn 1892: *Hylobates* sp.).
- Notes: Hepburn (1892) stated that in the *Hylobates* sp. specimen dissected by him the extensor carpi radialis longus runs from the lower part of the lateral supracondylar ridge and septum to the radial aspect of the base of metacarpal II and to the ulnar side of the base of metacarpal I, its muscular belly being small and its tendon long. Straus (1941a) stated that in *H. moloch* the extensor carpi radialis longus runs from both the lateral supracondylar ridge and the lateral epicondyle of the humerus to metacarpal I, while in *H. lar* and *H. pileatus* the muscle runs from the supracondylar ridge of the humerus to both metacarpals I and II. Deniker (1885) reported a fetal *Hylobates* sp. specimen in which the bellies of the extensor carpi radialis longus and of the extensor carpi radialis brevis are deeply blended, but their tendons are clearly distinct, inserting onto metacarpals II and III respectively. Kohlbrügge (1890–1892) dissected one *H. moloch* specimen, one *H. agilis* specimen, and one *H. syndactylus* specimen and stated that the extensor carpi radialis longus runs from the 'lateral condyle' of the humerus to the bases of metacarpals I and II, except in *H. syndactylus* and in one hand of *H. agilis*, in which the muscle goes to metacarpal II only. Duckworth (1904) reported one *H. muelleri* specimen in which the extensor carpi radialis longus originates from the lateral supracondylar line, while the extensor carpi radialis brevis originates from the condyle itself. Bojsen-Møller (1978) stated that in the two *H. lar* and the two *H. moloch* upper limbs dissected

by this author the extensor carpi radialis longus inserts on metacarpal II only. Michilsens et al. (2009) stated that in the 11 hylobatid specimens dissected by them (3 *H. lar*, 2 *H. pileatus*, 2 *H. moloch* and 4 *H. syndactylus*) the extensor carpi radialis longus runs from the lateral supracondylar ridge of the humerus to the bases of metacarpals I and II. In our *H. lar* specimen the extensor carpi radialis longus runs from the lateral supracondylar ridge of the humerus to metacarpals I and II, being largely distinct from the extensor carpi radialis brevis. In our *H. gabriellae* specimens VU HG1 and VU HG2 the extensor carpi radialis longus originates from the lateral supracondylar ridge of the humerus and only inserts onto the base of metacarpal II. In summary, in the 29 cases described in this paragraph the extensor carpi radialis longus goes to both metacarpals I and II in 18 (62%) of the 29 cases.

- Synonymy: Premier radial (Deniker 1885); extensor carpi radialis longior (Hartmann 1886, Hepburn 1892).

Extensor carpi radialis brevis (5.4 g; Figs. 27, 34)

- Usual attachments: From the lateral epicondyle and sometimes also from the lateral condyle/supracondylar ridge of the humerus to the base of metacarpal III and, often (seemingly usually also to the base of metacarpal II; occasionally the muscle may insert onto metacarpal II only.
- Usual innervation: Radial nerve (Kohlbrügge 1890–1892: *H. moloch, H. agilis, H. syndactylus*); posterior interosseous nerve (Hepburn 1892: *Hylobates* sp.).
- Notes: Straus (1941a) stated that in *H. moloch* and *H. lar* the extensor carpi radialis brevis runs from the lateral epicondyle of the humerus to metacarpal III, while in *H. pileatus* the muscle runs from the lateral epicondyle of the humerus to both metacarpals II and III. Deniker (1885) reported a fetal *Hylobates* sp. specimen in which the extensor carpi radialis brevis inserts onto metacarpal III. Kohlbrügge (1890–1892) stated that the extensor carpi radialis brevis runs from the 'lateral condyle' of the humerus to the bases of metacarpals II and III in one *H. moloch* specimen, in one *H. syndactylus* specimen and in one of the two hands of one *H. agilis* specimen, while in the other hand of this latter specimen the muscle goes to metacarpal III only. Michilsens et al. (2009) stated that within the 11 hylobatid specimens dissected by them (3 *H. lar*, 2 *H. pileatus*, 2 *H. moloch* and 4 *H. syndactylus*) the extensor carpi radialis brevis runs from the lateral epicondyle of the humerus and sometimes from the extensor digitorum to the bases of both metacarpals II and III, except in the three specimens of *H. lar*, in which the muscle goes to metacarpal III only and in one specimen of *H. syndactylus* in which it goes to metacarpal II only. In our *H. lar* specimen HU HL1 the extensor carpi radialis brevis runs from the lateral epicondyle of the humerus, where it partly blends with the extensor digitorum, to metacarpal III, while in our *H. gabriellae* specimen VU HG1 the extensor carpi radialis brevis runs from the lateral epicondyle of the humerus to the base of metacarpals II

and III. In summary, the extensor carpi radialis brevis goes to both metacarpals II and III in 13 (59%) of the 22 cases described in this paragraph.

- Synonymy: Deuxième radial (Deniker 1885); extensor carpi radialis brevior (Hartmann 1886, Hepburn 1892).

Brachioradialis (9.5 g; Figs. 27, 34)

- Usual attachments: From the shaft and distal end of the humerus to the shaft of the radius (usually not reaching the styloid process).
- Usual innervation: Radial nerve (Kohlbrügge 1890–1892: *H. moloch, H. agilis, H. syndactylus*; Hepburn 1892: *Hylobates* sp.).
- Notes: Hepburn (1892) stated that in the *Hylobates* sp. dissected by him the brachioradialis takes origin from the lateral supracondyloid ridge and septum, higher up than the other muscles arising from the same ridge, being intimately blended with the outer surface of the brachialis; its course lies along the radial border of the forearm to its point of insertion onto the anterior surface and outer border of the radius (for 2.5 inches; it fails to reach the styloid process by a distance of 2.5 inches). Straus (1941a) stated that the brachioradialis inserts at about middle of the radial shaft in *H. moloch*, and at the distal end of the middle 1/3 of this shaft in *H. lar*, the muscle fusing proximally with the brachialis in *H. moloch*. Barnard (1875) stated that the brachioradialis inserts onto the middle of the radius in *H. moloch*. Deniker (1885) reported a fetal *Hylobates sp.* specimen in which the insertion of the brachioradialis is more proximal than in modern humans, being to the middle of the radius (about 3 cm proximal to the styloid process). Kohlbrügge (1890–1892) reported *H. moloch, H. agilis* and *H. syndactylus* specimens in which the brachioradialis originates from the humerus (extending to the middle third of this bone in *H. syndactylus*) to the radius, not reaching the styloid process of the latter. Andrews and Groves (1976) wrote that the brachioradialis is massively developed in *Hylobates*, its origin extending up the humerus almost to the level of the distal end of the deltoid insertion; there is, however, no brachioradialis flange, in sharp contrast to the condition in the great apes. Michilsens et al. (2009) stated that within the 11 hylobatid specimens dissected by them (3 *H. lar*, 2 *H. pileatus*, 2 *H. moloch* and 4 *H. syndactylus*) the brachioradialis runs from the lateral supracondylar ridge of the humerus to the middle of radius, except in two specimens of *H. syndactylus*, in which it goes instead to the styloid process and in the other specimen of this species, in which it goes to the distal part of the radius (but, apparently, not to the styloid process). In our *H. lar* specimen HU HL1 the brachioradialis originates more distally than in the *Pan* and *Pongo* specimens dissected by us, being thus more similar to the condition in modern humans (from the 23 cm of the humerus, it originates directly only from the distal 5 cm, being deeply blended with the brachialis proximally); the muscle effectively inserts more proximally in hylobatids than in other hominoids, to about 4/5 of the total

length of the radius (i.e., to 5.5 cm proximal to the distal margin of the radius, which has a total length of 26 cm, so it does not insert onto the styloid process). In our *H. gabriellae* specimen VU HG1 the brachioradialis runs from the distal portion of the humerus to the radius, not reaching the styloid process. Payne (2001) dissected one hylobatid specimen in which the brachioradialis has a small ribbon-like tendon, which inserts further proximally than in the great apes analyzed by this author, onto the middle third of the lateral radial shaft. Payne (2001) suggested that the more proximal insertion of the brachioradialis onto the lateral radial shaft found in gibbons (compared with modern humans or orangutans) is probably related to the fact that in gibbons this muscle is required to function differently. That is, in modern humans and orangutans it inserts far from the joint axis, with relatively long fascicles, implying that this muscle is well suited for the production of high velocity movement over a wide range of joint positions, while the more proximal insertion of this muscle in gibbons indicates that it is well designed for forceful movements over a restricted area. According to Payne (2001) the configuration of the muscle in the gibbon is probably associated with a need for a "spurt" of elbow flexion at the end of the swing phase of ricochetal brachiation, as has been argued by Jungers and Stern (1984). Forelimb flexion brings the animals' centre of mass closer to the point of contact with the support, thus reducing deceleration due to inertia and maintaining rotational velocity. However, the hylobatid specimen dissected by Payne (2001) was found to have the shortest moment arm and thus leverage for this muscle, which may be due to its comparatively smaller origin on the lateral supra-condylar ridge of the humerus.

- Synonymy: Supinator longus (Barnard 1875, Deniker 1885, Kohlbrügge 1890–1892); radii longus (Hepburn 1892).

Supinator (5.9 g; Figs. 27, 34)

- Usual attachments: From the lateral epicondyle of the humerus (caput humerale, or superficiale) and usuallly also from the proximal part of the ulna (caput ulnare, or profundum) to the proximal radius.
- Usual innervation: Radial nerve (Kohlbrügge 1890–1892: *H. moloch*, *H. agilis*, *H. syndactylus*); posterior interosseous nerve (Hepburn 1892: *Hylobates* sp.); pierced and innervated by the posterior interosseous nerve (Straus 1941ab: *H. moloch*, *H. lar*, *H. pileatus*; our dissections: *H. gabriellae* specimen VU HG2).
- Notes: According to Straus (1941a) in hylobatids the supinator has ulnar and humeral heads and this was also the condition found in the three specimens dissected by Kohlbrügge (1890–1892: *H. moloch*, 1sp; *H. agilis*, 1 sp.; *H. syndactylus*, 1 sp.) and our *H. lar* specimen HU HL1 (in our *H. gabriellae* specimen VU HG1 the muscle only originated from the ulna) and the vast majority of authors claim that all the extant hominoid genera usually have both of these heads (e.g., Miller 1932; Jouffroy 1971; Lewis 1989; Gibbs

1999). However, Michilsens et al. (2009) stated that within the 11 hylobatids dissected by them (3 *H. lar*, 2 *H. pileatus*, 2 *H. moloch* and 4 *H. syndactylus*) the muscle originates from the humerus only, except in the three specimens of *H. lar*, in which the muscle also originates from the ulna. This contradicts the statements of Kohlbrügge (1890–1892) and Straus (1941a) according to which an ulnar head is also present in other hylobatid species such as *H. moloch* and *H. syndactylus* and Michilsens et al. (2009) stated that they did not found an ulnar head in six specimens belonging to the latter two species. In view of the data available, it is difficult to discern if this is the result of a true variation within this latter species and within the hylobatids in general, or if these contradictory statements are instead due to an error made by Michilsens et al. (2009).

- Synonymy: Supinator brevis (Kohlbrügge 1890–1892, Hepburn 1892).

Extensor carpi ulnaris (3.8 g; Figs. 27, 34)
- Usual attachments: From the lateral epicondyle of the humerus (caput humerale) and the ulna (caput ulnare) to metacarpal V.
- Usual innervation: Radial nerve (Kohlbrügge 1890–1892: *H. moloch*, *H. agilis*, *H. syndactylus*); posterior interosseous nerve (Hepburn 1892: *Hylobates* sp.).
- Notes: In hylobatids the extensor carpi ulnaris originates from the ulna and humerus, as noted by Kohlbrügge (1890–1892), Straus (1941a) and Michilsens et al. (2009) and corroborated by our dissections; in the specimens reported by others and dissected by us the muscle goes to metacarpal V only. The **anconeus** is usually not present as a distinct muscle in hylobatids, being undifferentiated from the extensor carpi ulnaris, as noted by Duckworth (1904) and corroborated by the descriptions of Kohlbrügge (1890–1892), Hepburn (1892), Payne (2001) and Michilsens et al. (2009) and by our dissections.
- Synonymy: Cubital postérieur (Deniker 1885).

Extensor digitorum (8.0 g; Figs. 27, 34)
- Usual attachments: From the lateral epicondyle of the humerus and often also from the radius and/or the ulna, to the middle phalanges and (via the extensor expansions) to the distal phalanges of digits 2, 3, 4 and 5; sometimes there is no insertion onto digit 5.
- Usual innervation: Radial nerve (Kohlbrügge 1890–1892: *H. moloch*, *H. agilis*, *H. syndactylus*); posterior interosseous nerve (Hepburn 1892: *Hylobates* sp.).
- Notes: Hepburn (1892) stated that in the *Hylobates* sp. specimen dissected by him the tendon of the extensor digitorum to digit 4 also sends a small slip to digit 5. Straus (1941a) stated that in hylobatids the muscle usually runs from the radial epicondyle of the humerus to digits 2–5, but that in one *H. moloch* there is also an origin from the ulna and in one *H. lar* specimen there is an insertion onto digits 2–4 only. Kaneff (1979) reported three *H. leucogenys* and two *H. syndactylus* specimens in which the extensor digitorum goes to digits 2–5. Deniker (1885) described a fetal *Hylobates* sp. specimen in which the extensor

digitorum goes from the radius, ulna and also humerus to the distal phalanges of digits 2–5. In the *Hylobates* sp. specimen illustrated in Fig. 52 of Hartmann (1886) the extensor digitorum goes to digits 2–5. Kohlbrügge (1890–1892) dissected one *H. moloch* specimen, one *H. agilis* specimen and one *H. syndactylus* specimen and stated that the muscle originates from the lateral epicondyle of the humerus and ulna, except in *H. syndactylus* in which there is no origin from the ulna, and inserts onto the middle and distal phalanges of digits 2–5 via the extensor expansions of these digits. Duckworth (1904) reported one *H. muelleri* specimen in which the extensor digitorum arises partially from the intermuscular septa and adjacent muscle sheaths; on the back of the hand the tendons spread out and are connected by an aponeurotic membrane. Payne (2001) reported one hylobatid specimen in which the muscle inserts onto the middle phalanx of the digits; according to this author full extension of the manual digits could not be achieved in the specimen. Michilsens et al. (2009) dissected 11 hylobatid specimens (3 *H. lar*, 2 *H. pileatus*, 2 *H. moloch* and 4 *H. syndactylus*) and stated that the extensor digitorum runs from the lateral epicondyle of the humerus to the middle phalanges of digits 2–4, except in the three specimens of *H. lar*, in which the muscle goes to digits 2–5 and not 2–4. In our *H. lar* specimen HU HL1 the extensor digitorum runs from the lateral epicondyle of the humerus and intermuscular septum and ulna to the middle phalanges of digits 2–5. In our *H. gabriellae* specimen VU HG1 the muscle runs from the lateral epicondyle of the humerus, antebrachial fascia, and intermuscular septum and ulna to the extensor expansions (and thus to both the middle and distal phalanges) of, seemingly, digits 2–4 only, while in our *H. gabriellae* specimen VU HG2 its bony origin is from the humerus only (not from the ulna or the radius) and its insertion is onto digits 2–5.

- Synonymy: Extensor digitorum communis (Barnard 1875, Deniker 1885, Straus 1941a,b); extensor communis digitorum (Hartmann 1886, Hepburn 1892); extensor digitorum sublimis (Kohlbrügge 1890–1892); extensor digitorum + part of extensor digiti minimi (Duckworth 1904).

Extensor digiti minimi (0.7 g; Figs. 27, 34)

- Usual attachments: From the lateral epicondyle of the humerus and/or the common extensor tendon and/or often also from the ulna, to the middle phalanx and (via the extensor expansion) to the distal phalanx of digit 5.
- Usual innervation: Radial nerve (Kohlbrügge 1890–1892: *H. moloch*, *H. agilis*, *H. syndactylus*); posterior interosseous nerve (Hepburn 1892: *Hylobates* sp.).
- Notes: In the hylobatid specimens dissected by Bischoff (1870), Deniker (1885), Hartmann (1886), Kohlbrügge (1890–1892), Hepburn (1892), Chapman (1900), Straus (1941a), Kaneff (1980a), Michilsens et al. (2009), and by us, the extensor digiti minimi inserts onto digit 5 only. In these specimens the origin is often from the humerus and/or common extensor tendon, except in one *H. moloch* specimen dissected by Straus (1941ab; in which the origin is

instead from the ulna only) in the three specimens of *H. lar* and in one of the four specimens of *H. syndactylus* dissected by Michilsens et al. (2009; in which the muscle has a bony origin from the humerus and from the middle portion of the ulna) in the fetal *Hylobates* sp. specimen dissected by Deniker (1885; in which the muscle originates from the ulna only) and in the three specimens (*H. moloch*, *H. agilis*, *H. syndactylus*) dissected by Kohlbrügge (1890–1892; in which the muscle originates from the ulna). Duckworth (1904) reported one *H. muelleri* specimen in which there is an 'extensor digiti minimi' constituted by two different muscles: 1) from the extensor digitorum a slip is given off to digit 5 (this corresponds to the part of the extensor digitorum *sensu* the present study); 2) an accessory muscle which is partly blended with the extensor digitorum and partly with the extensor carpi ulnaris (this corresponds to the extensor digiti minimi *sensu* the present study). The **extensor digiti quarti** is usually not present as a distinct muscle in hylobatids.

- Synonymy: Extenseur propre du petit doigt (Deniker 1885); extensor minimi digiti (Hartmann 1886, Hepburn 1892); extensor proprius minimi digiti (Chapman 1900); accessory muscle (Duckworth 1904); extensor digiti quarti et quinti proprius (Straus 1941a,b); extensor digitorum lateralis (Kaneff 1980a).

Extensor indicis (2.3 g; Figs. 27, 34)
- Usual attachments: From the ulna and often also from the interosseous membrane and less often also from radius and/or humerus, to digits 2, 3 and 4; occasionally there is no insertion onto one or more of these digits and/or there is an insertion onto digit 5.
- Usual innervation: Radial nerve (Kohlbrügge 1890–1892: *H. moloch*, *H. agilis*, *H. syndactylus*); posterior interosseous nerve (Hepburn 1892: *Hylobates* sp.; Aziz and Dunlap 1986: *H. syndactylus*).
- Notes: Within hylobatids, an insertion of the extensor indicis onto digits 2, 3 and 4 was described by Bischoff (1870), Kohlbrügge (1890–1892), Hepburn (1892), Chapman (1900), Jouffroy (1971), Kaneff (1980a,b), Aziz and Dunlap (1986) and Michilsens et al. (2009) and found by us, but in the *Hylobates* sp. fetus dissected by Deniker (1885) the insertion was onto digits 2–5, while Barnard (1875) described an insertion onto digit 3 only and Straus (1941a) onto digits 2 and 3 or 2, 3, and 4. According to the review of the literature done by Straus (1941a,b), in *Hylobates* an insertion onto digits 2, 3, 4 and 5 occurs in about 7.5% of the cases and onto digits 2, 3 and 4 occurs in about 92% of the cases. Regarding the origin of the extensor indicis, Deniker (1885) stated that in the fetal *Hylobates* sp. specimen dissected by him it is from the humerus, ulna and interosseous membrane, while Michilsens et al. (2009) stated that within the 11 hylobatids dissected by them (3 *H. lar*, 2 *H. pileatus*, 2 *H. moloch* and 4 *H. syndactylus*) it is from the distal 2/3 of the ulna, except in the three specimens of *H. lar* in which it is instead from the middle of the ulna. In our

H. lar specimen HU HL1 the extensor indicis runs from the distal 1/2 of the ulna (not from septum nor membrane) to the extensor expansions of digits 2, 3 and 4. On one side of our *H. gabriellae* specimen VU HG1 the muscle connects the interosseous membrane (we could not discern if it also originates from the ulna and/or radius) to the extensor expansions of digits 2, 3 and 4, while on the the other side it sends a tendon to digit 2, a tendon to digits 2 and 3, and a tendon to digits 3 and 4. In our *H. gabriellae* specimen VU HG2 it sends tendons to digits 2, 3 and 4. It should be noted that Duckworth (1904) described an 'extensor digiti medii et digiti annularis' in *H. muelleri*, but this structure actually corresponds to the portion of the extensor indicis that usually goes to digits 3 and 4 (in addition to digit 2) in hylobatids. The **extensor communis pollicis et indicis**, the **extensor digiti III proprius** and the **extensor brevis digitorum manus** (see, e.g., Diogo et al. 2009a) are usually not present as distinct muscles in hylobatids.

- Synonymy: Part or all of extensor profundus digitorum or extensor digitorum profundus (Barnard 1875, Kohlbrügge 1890–1892, Hepburn 1892, Straus 1941a,b and Kaneff 1980a); extenseur commun profond (Deniker 1885); it includes the extensor digiti medii et digiti annularis *sensu* Duckworth (1904); extensor digitorum profundus proprius or extensor indicis proprius + extensor medii digiti proprius + extensor digiti quarti proprius (Aziz and Dunlap 1986); extensor digitorum brevis (Michilsens et al. 2009).

Extensor pollicis longus (1.4 g; Figs. 27, 34)
- Usual attachments: From the ulna and interosseous membrane to the distal phalanx of the thumb and sometimes also to its proximal phalanx.
- Usual innervation: Radial nerve (Kohlbrügge 1890–1892: *H. moloch, H. agilis, H. syndactylus*); posterior interosseous nerve (Hepburn 1892: *Hylobates* sp.).
- Notes: Hepburn (1892) stated that in the *Hylobates* sp. specimen dissected by him the extensor pollicis longus attaches to the proximal phalanx of the thumb. Straus (1941a,b) stated that in hylobatids the muscle usually goes to the distal phalanx of digit 1, but that in *H. pileatus* it also goes to the proximal phalanx of this digit. Deniker (1885) reported one fetal *Hylobates* sp. specimen in which the extensor pollicis longus goes to the distal phalanx of digit 1. Kohlbrügge (1890–1892) dissected *H. moloch, H. agilis* and *H. syndactylus* specimens and stated that the muscle goes to the proximal and distal phalanges of the thumb. Chapman (1900) also stated that in the *H. moloch* specimen dissected by him the extensor pollicis longus apparently acts on both the proximal and distal phalanges of the pollex. Michilsens et al. (2009) dissected 11 hylobatids (3 *H. lar*, 2 *H. pileatus*, 2 *H. moloch* and 4 *H. syndactylus*) and stated that the extensor pollicis longus runs from the proximal 1/4 of the ulna to the terminal phalanx of the pollex. In our *H. lar* specimen HU HL1 the extensor pollicis longus runs from the ulna and interosseous membrane to the distal phalanx of digit 1, being

well differentiated from the extensor indicis. In our *H. gabriellae* specimen VU HG1 the extensor pollicis longus runs from the interosseous membrane (we could not discern if it was also originated from the ulna and/or radius) to the distal phalanx of digit 1, being somewhat blended with, but still clearly distinct from, the extensor indicis. In our *H. gabriellae* specimen VU HG2 the extensor pollicis longus inserts onto the distal phalanx of the thumb.

- Synonymy: Extensor secundi internodii pollicis (Hartmann 1886, Chapman 1900, Hepburn 1892); part or totality of extensor profundus digitorum or extensor digitorum profundus (Straus 1941a,b).

Abductor pollicis longus (3.4 g; Figs. 27, 34)
- Usual attachments: See notes below.
- Usual innervation: Radial nerve (Kohlbrügge 1890–1892: *H. moloch, H. agilis, H. syndactylus*); posterior interosseous nerve (Hepburn 1892: *Hylobates* sp.).
- Notes: Apart from modern humans, within all the primate specimens dissected by us a distinct extensor pollicis brevis displaying a distinct muscular belly that is only partially blended, proximally, with the belly of the abductor pollicis longus, is only present in hylobatids (e.g., described by Bischoff 1870, Kohlbrügge 1890–1892, Duckworth 1904, and Michilsens et al. 2009, although Deniker 1885 stated that he did not found a distinct extensor pollicis brevis in the *Hylobates* sp. fetus dissected by him). It should be noted that some authors described an 'extensor pollicis brevis' and an 'abductor pollicis longus' in some primate taxa other than hylobatids and modern humans. The name 'extensor pollicis brevis' has for instance often been used in descriptions of gorillas (e.g., Hepburn 1892; Straus 1941a,b) because in these primates a tendon of the abductor pollicis longus (*sensu* the present study) often inserts onto the proximal phalanx of the thumb, i.e., to the typical insertion point of the extensor pollicis brevis of modern humans. However, as stressed by authors such as Deniker (1885), Kaneff (1979, 1980a,b) and Aziz and Dunlap (1986) and corroborated by our dissections in gorillas there is actually usually a single fleshy belly of the abductor pollicis longus that then gives rise to the so-called 'tendons of the extensor pollicis brevis and of the abductor pollicis longus', as is usually also the case in *Pongo* and *Pan*. That is, contrary to *Homo* and hylobatids, in *Pongo, Pan* and *Gorilla* the extensor pollicis brevis is usually not present as a separate muscle. Hepburn (1892) reported one *Hylobates* sp. specimen in which the abductor pollicis longus inserts onto the sesamoid bone ('prepollex') while the extensor pollicis brevis inserts onto the base of metacarpal I. Straus (1941a,b, 1942b) stated that in hylobatids the complex formed by the abductor pollicis longus and extensor pollicis brevis usually originates from both the radius and ulna, but that in *H. moloch* it originates from the ulna only; insertion of the complex is onto the 'pre-pollex', trapezium and metacarpal I in *H. pileatus*, onto the 'pre-pollex' and metacarpal I in *H. moloch*, and onto the trapezium

and metacarpal I in *H. pileatus*. Deniker (1885) reported one fetal *Hylobates* sp. specimen fetus in which the 'abductor pollicis longus' (abductor pollicis longus + extensor pollicis brevis *sensu* the present study) originates from the radius and has a tendon to the base of metacarpal I and a tendon to the proximal phalanx of the thumb; he stated that in the *H. moloch* specimen dissected by Bischoff (1870) there is a distinct extensor pollicis brevis (no reference to its insertion) and a distinct abductor pollicis longus, the latter inserting onto the base of metacarpal I and onto the trapezium. Hartmann (1886) stated that the abductor pollicis longus and extensor pollicis brevis send tendons to metacarpal I and the trapezium in hylobatids. Kohlbrügge (1890–1892) dissected *H. moloch*, *H. agilis* and *H. syndactylus* specimens and stated that the abductor pollicis longus runs mainly from the ulna and radius to the 'prepollex' sesamoid bone and to the trapezium, while the extensor pollicis brevis runs mainly from the radius and ulna (but not from the ulna in *H. moloch*) to the metacarpal I. Chapman (1900) reported one *H. moloch* specimen in which the 'abductor pollicis longus' (abductor pollicis longus + extensor pollicis brevis *sensu* the present study) gives rise to two tendons, one to the metacarpal I, the other to the trapezium. Duckworth (1904) stated that in the *H. muelleri* specimen dissected by him the adductor pollicis longus arises from the radius, ulna and interosseous membrane, while the extensor pollicis brevis arises from the radius, ulna and interosseous membrane. Wood Jones (1920) stated that in *Hylobates* the 'abductor pollicis longus' (abductor pollicis longus + extensor pollicis brevis *sensu* the present study) usually inserts onto the carpus, mainly onto the trapezium, and sometimes also onto the adjacent sesamoid bone. Lorenz (1974) asserted that in *Hylobates* the abductor pollicis longus inserts onto the carpal region and thus abducts the hand as a whole, while the extensor pollicis brevis is weak and inserts onto metacarpal I. Michilsens et al. (2009) stated that in the 11 hylobatids dissected by them (3 *H. lar*, 2 *H. pileatus*, 2 *H. moloch* and 4 *H. syndactylus*) the abductor pollicis longus runs from the proximal 1/3 of the ulna and radius to the medial side of the base of metacarpal I, except in one specimen of *H. syndactylus* in which the muscle connects the proximal half of the ulna and radius to the trapezium (2 tendons), and other specimen of the same species, in which the insertion is onto the trapezium but by a single tendon; the extensor pollicis brevis runs from the proximal medial part of the radius and the interosseous membrane to the dorsal side of metacarpal I, except in the three specimens of *H. lar* in which the muscle originates from the middle 1/3 of the radius, in these three specimens and one specimen of *H. syndactylus* in which the muscle inserts onto the medial part of the base of metacarpal I, and in another specimen of *H. syndactylus*, in which the muscle connects the proximal 1/2 of the radius and also the proximal ulna to the scaphoid and trapezium (although, as they recognized, this latter insertion could be of the abductor pollicis longus and not of the extensor pollicis brevis).

In our *H. lar* specimen HU HL1 the abductor pollicis longus runs from the radius, ulna and interosseous membrane to a small sesamoid bone that lies just lateral (radial to) to the trapezium; the extensor pollicis brevis is a distinct muscle with a separate muscular belly, which is only deeply blended with the belly of the abductor pollicis longus proximally and which originates together with the abductor pollicis longus from the radius and interosseous membrane and possible also the ulna, and inserts onto the trapezius and possibly the base of metacarpal I. In our *H. gabriellae* specimen VU HG1 the abductor pollicis longus runs from the radius, ulna and interosseous membrane to the sesamoid bone and/or trapezium, while the extensor pollicis brevis runs from the interosseous membrane and possibly the radius (but not the ulna) to the base of metacarpal I; the tendons of these two muscles are completely separated, but their fleshy bellies are blended proximally, seemingly more blended than they usually are in modern humans.

- Synonymy: Extensor ossi metacarpi pollicis (Hepburn 1892, Chapman 1900, Duckworth 1904); abductor longus pollicis (Hartmann 1886).

Extensor pollicis brevis (1.6 g; Figs. 27, 34)
- Usual attachments: See notes about abductor pollicis longus above.
- Usual innervation: Radial nerve (Kohlbrügge 1890–1892: *H. moloch*, *H. agilis*, *H. syndactylus*); posterior interosseous nerve (Hepburn 1892: *Hylobates* sp.).
- Notes: See notes about abductor pollicis longus, above.
- Synonymy: Extensor primi internodii pollicis (Hartmann 1886).

CHAPTER 5

Trunk and Back Musculature

Obliquus capitis inferior (Fig. 17)
- Usual attachments: To our knowledge, there are no detailed descriptions of this muscle in hylobatids. In our *H. gabriellae* specimen VU HG2 the muscle runs from the spinous process of C2 to the transverse process of C1.
- Usual innervation: Data are not available.

Obliquus capitis superior (Fig. 17)
- Usual attachments: To our knowledge, there are no detailed descriptions of this muscle in hylobatids. In our *H. gabriellae* specimen VU HG2 the muscle runs from the transverse process of C1 to the inferior nuchal line of the occipital bone.
- Usual innervation: Data are not available.

Rectus capitis anterior
- Usual attachments: To our knowledge, there are no detailed descriptions of this muscle in hylobatids. In our *H. gabriellae* specimen VU HG2 the muscle runs from the lateral portion of the atlas to the basilar part of the occipital bone.
- Usual innervation: Data are not available.

Rectus capitis lateralis
- Usual attachments: To our knowledge, there are no detailed descriptions of this muscle in hylobatids. In our *H. gabriellae* specimen VU HG2 the muscle runs from the transverse process of atlas to the jugular process of the occipital bone.
- Usual innervation: Data are not available.

Rectus capitis posterior major (1.1 g)
- Usual attachments: To our knowledge, there are no detailed descriptions of this muscle in hylobatids. In our *H. gabriellae* specimen VU HG2 the muscle runs from the spinous process of C2 to the inferior nuchal line of the occipital bone.
- Usual innervation: Data are not available.

Rectus capitis posterior minor (0.5 g)
- Usual attachments: To our knowledge, there are no detailed descriptions of this muscle in hylobatids. In our *H. gabriellae* specimen VU HG2 the muscle runs from the posterior tubercle of C1 to the inferior nuchal line of the occipital bone.
- Usual innervation: Data are not available.

Longus capitis (Figs. 5, 6)
- Usual attachments: To our knowledge, there are no detailed descriptions of this muscle in hylobatids.
- Usual innervation: Data are not available.

Longus colli (Fig. 6)
- Usual attachments: To our knowledge, there are no detailed descriptions of this muscle in hylobatids.
- Usual innervation: Data are not available.

Scalenus anterior (Fig. 5)
- Usual attachments: Gibbs (1999) stated that in gibbons the origin of this muscle is described as from a variable number of vertebrae between C3 and C6, while Stewart (1936) wrote that in *H. lar* the scalenus anterior runs from the transverse processes of C4, C5 and C6 to the first rib. In our *H. gabriellae* specimens VU HG1 and VU HG2 the muscle is similar to that usually found in modern humans.
- Usual innervation: Data are not available.
- Notes: Stewart (1936) stated that hylobatids and other apes usually have a scalenius anterior and a 'scalenus medius' (which corresponds to the scalenus medius + scalenus posterior *sensu* the present study). According to him the scalenius anterior usually inserts onto the first rib in primates, its origin being limited to C5 and C6 in 'lower primates', but in hominoids the insertion included some of the upper cervical vertebrae. Also according to him, within primates the 'scalenus medius' (i.e., medius + posterior *sensu* the present study) shows a progressive decrease in the number of ribs supplying attachment, usually inserting in hominoids upon the first rib, but occasionally also onto the second rib.

Scalenus medius (Fig. 5)
- Usual attachments: Deniker (1885) stated that in the fetal *Hylobates* sp. specimen dissected by him there is no distinct 'scalenus medius' (which corresponds to the scalenus medius + scalenus posterior *sensu* the present study). However, Stewart (1936) wrote that in *H. lar* the 'scalenus medius' (scalenus medius + scalenus posterior *sensu* the present work) is a stout muscle running from the transverse processes of C2–C6 to the first rib.
- Usual innervation: Data are not available.
- Notes: See notes about scalenus medius above.

Scalenus posterior (Fig. 5)
- Usual attachments: To our knowledge, there are no detailed descriptions of this muscle in hylobatids.
- Usual innervation: Data are not available.
- Notes: See notes about scalenus medius above. To our knowledge, the **scalenus minimus** was not reported in hylobatids.

Levatores costarum (Fig. 16)
- Usual attachments: To our knowledge, there are no detailed descriptions of these muscles in hylobatids. In our *H. gabriellae* specimen VU HG2 the muscles run mainly from the transverse processes of the thoracic vertebrae to the ribs that lie posteriorly to these vertebrae.
- Usual innervation: Data are not available.

Intercostales externi (Figs. 16, 17)
- Usual attachments: To our knowledge, there are no detailed descriptions of these muscles in hylobatids. In our *H. gabriellae* specimen VU HG2 the muscles connect the adjacent margins of each pair of ribs.
- Usual innervation: Data are not available.

Intercostales interni
- Usual attachments: To our knowledge, there are no detailed descriptions of these muscles in hylobatids. In our *H. gabriellae* specimen VU HG2 the muscles connect the adjacent margins of each pair of ribs.
- Usual innervation: Data are not available.

Transversus thoracis
- Usual attachments: To our knowledge, there are no detailed descriptions of this muscle in hylobatids.
- Usual innervation: Data are not available.

Splenius capitis (5.8 g; Figs. 6, 8, 9, 10, 16)
- Usual attachments: Stewart (1936) stated that in *H. lar* the splenius capitis originates from the spine of T1 up to within 1cm of the inion, and inserts onto the superior nuchal line of the occiput. In the *H. lar* female dissected by Donisch (1973) the splenius capitis originates from the lower portion of the ligamentum nuchae to T1, and inserts onto the superior nuchal line. In our *H. gabriellae* specimen VU HG2 the muscle originates from the spinous processes of C5 to T3 and inserts onto the external margin of the superior nuchal line and also onto the mastoid process.
- Usual innervation: Data are not available.
- Synonymy: Part of splenius cervicus et capitis (Gibbs 1999).

Splenius cervicis (3.2 g; Fig. 17)
- Usual attachments: In our *H. gabriellae* specimen VU HG2 the muscle runs from the spinous processes of T3–T4 to the transverse processes of C1–C4.
- Usual innervation: Data are not available.

Serratus posterior superior (0.8 g; Fig. 16)
- Usual attachments: To our knowledge, there are no detailed descriptions of this muscle in hylobatids. In our *H. gabriellae* specimen VU HG2 the muscle runs from the spinous processes of C4-T1 to the second, third and fourth ribs.
- Usual innervation: Data are not available.

Serratus posterior inferior (1.4 g; Fig. 15)
- Usual attachments: To our knowledge, there are no detailed descriptions of this muscle in hylobatids. In our *H. gabriellae* specimen VU HG2 the muscle runs from the spinous processes of T10-L2 to ribs 10, 11, 12, 13 and 14.
- Usual innervation: Data are not available.

Iliocostalis (Fig. 15)
- Usual attachments: To our knowledge, there are no detailed descriptions of this muscle in hylobatids. In our *H. gabriellae* specimen VU HG2 the muscle runs from the iliac crest to all ribs.
- Usual innervation: Data are not available.
- Notes: The iliocostalis, longissimus and spinalis form the '**erector spinae**' (see Fig. 15).

Longissimus (54.7 g; Figs. 15, 16, 17)
- Usual attachments: To our knowledge, there are no detailed descriptions of this muscle in hylobatids. In our *H. gabriellae* specimen VU HG2 the muscle mainly runs from the inner side of the iliac crest, the sacrum, and the spinous processes of the lumbar vertebrae, to the ribs, the posterior tubercle of the transverse processes of C2–C6 and to the mastoid process.
- Usual innervation: Data are not available.
- Notes: The iliocostalis, longissimus and spinalis form the '**erector spinae**' (see Fig. 15). In modern humans, gorillas and a few other primates the **atlantomastoideus** might be occasionally present as a distinct muscle, running from the atlas to the mastoid process, but it was missing in the hylobatids dissected by us and to our knowledge it was not reported in any other hylobatids that have been dissected by other authors.

Spinalis (Figs. 15, 16)
- Usual attachments: To our knowledge, there are no detailed descriptions of this muscle in hylobatids. In our *H. gabriellae* specimen VU HG2 the muscle mainly connects the spinous processes of different vertebrae.
- Usual innervation: Data are not available.
- Notes: The iliocostalis, longissimus and spinalis form the '**erector spinae**' (see Fig. 15).

Semispinalis thoracis (Fig. 16)
- Usual attachments: To our knowledge, there are no detailed descriptions of this muscle in hylobatids. In our *H. gabriellae* specimen VU HG2 the muscle runs from the transverse processes of the last seven thoracic vertebrae to the

spinous processes of the first six thoracic vertebrate and of the last two cervical vertebrae.
- Usual innervation: Data are not available.

Semispinalis cervicis
- Usual attachments: To our knowledge, there are no detailed descriptions of this muscle in hylobatids. In our *H. gabriellae* specimen VU HG2 the muscle runs from the transverse processes of the first six thoracic vertebrae to the spinous processes of C2–C5.
- Usual innervation: Data are not available.

Semispinalis capitis (7.3 g; Figs. 16, 17)
- Usual attachments: To our knowledge, there are no detailed descriptions of this muscle in hylobatids. In our *H. gabriellae* specimen VU HG2 the muscle runs from the transverse processes of the first six thoracic vertebrae and of the last fourth cervical vertebrae to the occipital bone above the inferior nuchal line.
- Usual innervation: Data are not available.

Multifidus (Fig. 17)
- Usual attachments: To our knowledge, there are no detailed descriptions of this muscle in hylobatids.
- Usual innervation: Data are not available.

Rotatores
- Usual attachments: To our knowledge, there are no detailed descriptions of this muscle in hylobatids.
- Usual innervation: Data are not available.
- Synonymy: Rotatores breves and longi (Gibbs 1999).

Interspinalis
- Usual attachments: To our knowledge, there are no detailed descriptions of this muscle in hylobatids.
- Usual innervation: Data are not available.

Intertransversarii
- Usual attachments: To our knowledge, there are no detailed descriptions of this muscle in hylobatids.
- Usual innervation: Data are not available.

CHAPTER 6

Diaphragmatic and Abdominal Musculature

Diaphragma
- Usual attachments: According to Juraniec (1972) and Gibbs (1999) in *H. syndactylus* and in other hominoids such as chimpanzees and modern humans the position of the aortic aperture is higher than in 'lower catarrhines', the aperture being shorter by two-thirds of a vertebra in modern humans than in chimpanzees and in *H. syndactylus*. The division of the central tendon into folia is less distinct in modern humans and chimpanzees than in *H. syndactylus*, but the overall shape of this tendon is similar in these three taxa, with lateral and anterior leaflets of approximately the same size; in 1/2 *H. syndactylus* muscle fibres were observed running obliquely through the anterior leaflet and also at the base and centre of the right folia (Juraniec and Szostakiewicz-Sawicka 1968; Gibbs 1999). The oesophageal hiatus is situated at the level of T9 to T11 in *H. syndactylus*, being elliptical in shape and being formed by the splitting of the medial fibres of the right crus; the opposing ends of the oesophageal and aortic hiatuses are separated by half a vertebra (Juraniec 1972; Gibbs 1999). Regarding the type of fibre crossing at the oesophageal hiatus, Juraniec (1972) found the muscle fibres of the right crus crossing the fibres of the left crus (mingling with them) on one side of a *H. syndactylus* specimen; on the other side of this specimen a tendinous strand at the site of the crossing of the fibres was present on both the thoracic and the abdominal sides.
- Usual innervation: Data are not available.

Rectus abdominis
- Usual attachments: To our knowledge, there are no detailed descriptions of this muscle in hylobatids.
- Usual innervation: Data are not available.
- Notes: In gibbons the superficial rectus sheath is formed by the external and internal oblique and the deep sheath by the transversus abdominis alone (Lunn

1949; Gibbs 1999). To our knowledge, the **pyramidalis**, the **supracostalis** and the **tensor linea semilunaris** were not reported in hylobatids.

Cremaster

- Usual attachments: With the exception of a single gibbon reported by Mijsberg (1923) in hylobatids the muscle usually originates from the internal abdominal oblique (Miller 1947; Lunn 1949; Hill and Kanagasuntheram 1959). The cremaster usually contains a contribution from the transverses abdominis (Mijsberg 1923; Miller 1947; Lunn 1949; Hill and Kanagasuntheram 1959).
- Usual innervation: Data are not available.

Obliquus externus abdominis

- Usual attachments: According to authors such as Lunn (1948) and Miller (1947) there is no true inguinal ligament in any ape, only a series of tendinous arches over the sartorius and the femoral vessels and nerves, merging with the fascial lata. According to Winckler (1950) in hylobatids the obliquus externus abdominis may reach as far superiorly as the fourth rib.
- Usual innervation: Data are not available.

Obliquus internus abdominis

- Usual attachments: In hylobatids the obliquus internus abdominis muscle usually forms the conjoined tendon with the transversus abdominis, and the former is often described with the latter (Miller 1947; Lunn 1949; Hill and Kanagasuntheram 1959).
- Usual innervation: Data are not available.

Transversus abdominis

- Usual attachments: In hylobatids this muscle usually forms the posterior layer of the rectus sheath (Walmsley 1937; Lunn 1949). The fibres of the transversus abdominis decussate in the linea alba and the aponeurotic part forms, along with the internal oblique, the conjoined tendon, which inserts onto the superior pubic surface in the region of the pubic crest and in hylobatids it may also attach to the anterior wall of the rectus sheath above the pubis (Miller 1947; Lunn 1949; Hill and Kanagasuntheram 1959).
- Usual innervation: Data are not available.

Quadratus lumborum (Fig. 37)

- Usual attachments: To our knowledge, there are no detailed descriptions of this muscle in hylobatids. In our *H. gabriellae* specimen VU HG2 the muscle runs from the last rib to the costal processes of the lumbar vertebrae and to the iliac crest.
- Usual innervation: Data are not available.

CHAPTER 7

Perineal, Coccygeal and Anal Musculature

Coccygeus
- Usual attachments: In gibbons the muscle runs from the ischial spine to the side of the coccyx and to the anococcygeal raphe. The muscle is mostly tendinous with few muscle fibres, and it is reported to be more extensively developed than in other apes (Elftman 1932). According to Elftman (1932) in hylobatids there is no trace of the **flexor caudae** and the **iliococcygeus** is replaced by fascia.
- Usual innervation: Data are not available.

Levator ani
- Usual attachments: To our knowledge, there are no detailed descriptions of this muscle in hylobatids.
- Usual innervation: Data are not available.

Pubovesicalis
- Usual attachments: To our knowledge, there are no detailed descriptions of this muscle in hylobatids.
- Usual innervation: Data are not available.
- Synonymy: Ligamentum puboprostaticum or puboampullaris (Gibbs 1999).

Pubococcygeus
- Usual attachments: In hylobatids the pubococcygeus usually unites at the midline with its counterpart where it inserts onto the rectal wall, the tip of the coccyx and into the anococcygeal raphe (Thompson 1901; Smith 1923; Elftman 1932).
- Usual innervation: Data are not available.

Puborectalis
- Usual attachments: To our knowledge there are no detailed descriptions of this muscle in hylobatids.
- Usual innervation: Data are not available.

Sphincter ani externus
- Usual attachments: The sphincter encircles the anus, some of its deeper fibres radiating into the raphe of the bulbospongiousus (Elftman 1932).
- Usual innervation: Data are not available.

Bulbospongiosus
- Usual attachments: In hylobatids the muscle usually originates from the median raphe of the penile bulb and from the ischium, which is more extensive in gibbons than in African apes; the muscle surrounds the bulb and the corpora of the penis and fibres of the external anal sphincter merge with the superior fibres (Elftman 1932).
- Usual innervation: Data are not available.
- Synonymy: Bulbocavernosus (Gibbs 1999).

Ischiocavernosus
- Usual attachments: To our knowledge there are no detailed descriptions of this muscle in hylobatids.
- Usual innervation: Data are not available.

Sphincter urethrae
- Usual attachments: To our knowledge there are no detailed descriptions of this muscle in hylobatids.
- Usual innervation: Data are not available.

Transversus perinei profundus
- Usual attachments: To our knowledge there are no detailed descriptions of this muscle in hylobatids.
- Usual innervation: Data are not available.

Transversus perinei superficialis
- Usual attachments: To our knowledge there are no detailed descriptions of this muscle in hylobatids.
- Usual innervation: Data are not available.

Rectococcygeus
- Usual attachments: To our knowledge there are no detailed descriptions of this muscle in hylobatids.
- Usual innervation: Data are not available.

Rectourethralis
- Usual attachments: To our knowledge there are no detailed descriptions of this muscle in hylobatids.
- Usual innervation: Data are not available.
- Notes: To our knowledge the muscles **rectovesicalis, regionis analis, regionis urogenitalis, compressor urethrae, sphincter urethrovaginalis, sphincter pyloricus, suspensori duodeni, sphincter ani internus, sphincter ductus choledochi, sphincter ampullae, detrusor vesicae, trigoni vesicae,**

vesicoprostaticus, **vesicovaginalis**, **puboprostaticus** and **rectouterinus** have not been described in detail in hylobatids. We could also not examine these muscles in detail in the hylobatids dissected by us because this region had been damaged prior to our dissections.

CHAPTER 8

Pelvic and Lower Limb Musculature

Iliacus (16.5 g)
- Usual attachments: From the iliac fossa (Hepburn 1892) and the anterior surface of the ilium (Sigmon 1974). It inserts, in combination with the psoas major, onto the medial aspect of the lesser trochanter and more distally onto the adjacent femoral shaft (Beddard 1893, Sigmon 1974).
- Usual innervation: Branches of femoral nerve (Sigmon 1974).

Psoas major (16.2 g; Figs. 37, 41, 42, 43)
- Usual attachments: From the lateral surfaces of the bodies and the costal processes of the lumbar vertebrae, extending proximally to T12 and distally to S1 and from the intervening intervertebral discs (Hepburn 1892; Sigmon 1974). According to Hepburn (1892) and Kohlbrügge (1890–1892) it also has an origin from the lateral side of the tendon of origin of the rectus femoris. The psoas major merges with the iliacus and inserts as a combined iliopsoas tendon onto the lesser trochanter and more distally onto the adjacent femoral shaft (Hepburn 1892; Beddard 1893; Sigmon 1974).
- Usual innervation: Femoral nerve and first two or three lumbar nerves (Hepburn 1892; Sigmon 1974).

Psoas minor (Figs. 37, 41, 42, 43)
- Usual attachments: From the anterolateral surface of L1 (Kohlbrügge 1890–1892; Hepburn 1892; Sigmon 1979). An origin from the last thoracic vertebra is present in half of the hylobatids reported by Sigmon (1974) and an origin form L2 may also be present (Kohlbrügge 1890–1892; Hepburn 1892). In a single hylobatid specimen dissected by Kohlbrügge (1890–1892) the origin extends further inferiorly to L3. According to Sigmon (1974) in hylobatids the psoas minor also takes origin from the intervertebral discs and lies on the anteromedial surface of the psoas major, being fused with the latter muscle in two-thirds of the hylobatids reported by this author and in all the three

hylobatids in which we examined this feature in detail (*H. lar* specimens HU HL1 and GWU HL1, *H. gabriellae* specimen VU HG2). The psoas minor inserts onto the iliopubic eminence and the pectineal line (Hepburn 1892; Sigmon 1974).

- Usual innervation: First lumbar nerve (Hepburn 1892; Sigmon 1974); it may also be innervated by the twelfth thoracic nerve according to Sigmon (1974).

Gluteus maximus (27.9 g; Fig. 37)
- Usual innervation: It originates from the posterior iliac crest, the thoracolumbar fascia, sacrum, coccyx, sacrotuberal ligament, the fascia over gluteus medius, and the ischial tuberosity (Kohlbrügge 1890–1892; Hepburn 1892; Beddard 1893; Ranke 1897; Van den Broek 1914; Mysberg 1917; Miller 1945; Robinson et al. 1972; Sigmon 1974; Zihlman and Brunker 1979; Brown 1983; Hamada 1985) as well as from the posterior superior iliac spine where it shares its origin with that of the long head of biceps femoris (Stern 1972). It inserts onto the iliotibial tract (when present), the posterolateral aspect of the femur in the region of the gluteal tuberosity and the aponeurosis of the vastus lateralis (Appleton and Ghey 1929; Sigmon 1974; Brown 1983; Hamada 1985). The insertion of the muscle is generally more distal on the femur in hylobatids and other apes than in modern humans, extending almost two-thirds down the femur in gibbons (Stern 1972, Hamada 1985; Stern 1972; Sigmon 1974) and down the proximal four-fifths of the femoral shaft in siamangs (Sigmon 1974). The gluteus maximus is a flat muscle with its proximal portion being thinner than the distal portion (Hepburn 1892; Sigmon 1974; Zihlman and Brunker 1979). The muscle may be subdivided into as many as three parts (Hamada 1985).
- Usual innervation: Inferior gluteal nerve (our dissections: *H. lar* specimens HU HL1 and GWU HL1, *H. gabriellae* specimen VU HG2).

Gluteus medius (25.5 g; Figs. 38, 47, 48)
- Usual attachments: From the lateral surface of the ilium and the gluteal fascia (Beddard 1893) to the lateral aspect of the greater trochanter (Sigmon 1974), always fusing with the piriformis at its insertion (Beddard 1893) and being smaller than the gluteus maximus (Sigmon 1974).
- Usual innervation: Superior gluteal nerve (Sigmon 1974).

Gluteus minimus (2.3 g; Figs. 38, 47, 48)
- Usual attachments: It originates from the dorsolateral surface of the ilium extending to the ischial spine and from just distal to the anterior superior iliac spine, towards the acetabulum (Sigmon 1974). The muscle is triangular or fan-shaped (Sigmon 1974) and it is usually larger in African apes than in the Asian apes (Hepburn 1892). It inserts onto the anterior aspect of the greater trochanter (Hepburn 1892; Sigmon 1974).

- Usual innervation: Superior gluteal nerve (Sigmon 1974; our dissections: *H. lar* specimens HU HL1 and GWU HL1, *H. gabriellae* specimen VU HG2).

Scansorius (3.3 g; Fig. 38, 47)
- Usual attachments: The scansorius is said to be occasionally present in hylobatids (Hepburn 1892; Beddard 1893; Blake 1976); when it is present it is a flat triangular muscle that is fused with the gluteus minimus (Beddard 1893; Sigmon 1974).
- Usual innervation: Data are not available.
- Notes: Within the three hylobatids in which we examined this region in detail (*H. lar* specimens HU HL1 and GWU HL1, *H. gabriellae* specimen VU HG2), the scansorius is clearly present in two specimens (*H. lar* specimen GWU HL1, *H. gabriellae* specimen VU HG2; see Fig. 38, 47).

Ischiofemoralis (17.1 g; Fig. 37)
- Usual attachments: The ischiofemoralis is often considered to be part of the biceps femoris. However, in at least some hylobatids the ischiofemoralis does seem to represent a distinct muscle (e.g., in our *H. gabriellae* specimen VU HG2: Fig. 37), often originating from the ischial tuberosity (Brown 1983); its most cranial fibres (which are fused to the gluteus maximus) insert directly onto the proximal femoral shaft, while its most distal fibres penetrate the tensor fascia lata and insert onto the aponeurosis of the vastus lateralis (Brown 1983). The ischiofemoralis has a common tendinous aponeurosis with the biceps femoris and the semitendinosus (Brown 1983).
- Usual innervation: Branches of inferior gluteal nerve (Brown 1983).

Gemellus superior (0.1 g; Figs. 39, 48)
- Usual attachments: Sigmon (1974) stated that this muscle is absent in hylobatids, although in one case a small fasciculus can be raised from the obturator internus (Hepburn 1892). In the three hylobatids in which we examined this region in detail (*H. lar* specimens HU HL1 and GWU HL1, *H. gabriellae* specimen VU HG2) the gemellus superior is a distinct narrow bundle that originates from the region of ischial spine and inserts onto the trochanteric fossa, along with the tendon of the obturator internus.
- Usual innervation: Branches of sacral plexus (our dissections: *H. lar* specimens HU HL1 and GWU HL1, *H. gabriellae* specimen VU HG2).

Gemellus inferior (0.3 g; Figs. 39, 48)
- Usual attachments: It originates from the ischial tuberosity (Hepburn 1892) and the area just above it (Sigmon 1974). Close to its insertion into the trochanteric fossa (Beddard 1893; Sigmon 1974) it fuses with the inferior border of the obturator internus (Hepburn 1892; Sigmon 1974).
- Usual innervation: Branches of sacral plexus (Sigmon 1974).

Obturatorius externus (5.5 g; Fig. 42)

- Usual attachments: It runs from the external surface of the medial bony margin of the obturator foramen and from the obturator membrane (Sigmon 1974). It fuses with the obturator internus (Hepburn 1892) before it inserts with it into the trochanteric fossa (Hepburn 1892; Sigmon 1974).
- Usual innervation: Obturator nerve (Sigmon 1974).
- Synonymy: Obturator internus (Gibbs 1999).

Obturatorius internus (3.0 g; Figs. 39, 48)

- Usual attachments: It originates from the margin of the obturator foramen and from the obturator membrane (Sigmon 1974) to form a flat muscle that runs through the lesser sciatic foramen; there are no ridges and grooves where the tendon takes a right-angled turn into the buttock (Hepburn 1892). The insertion is in common with the inferior gemellus onto the trochanteric fossa (Beddard 1893; Sigmon 1974).
- Usual innervation: Sacral nerves (Sigmon 1974).
- Synonymy: Obturator externus (Gibbs 1999).

Piriformis (5.0 g; Figs. 38, 47, 48)

- Usual attachments: It originates by slips from the anterolateral surface of the distal half of the sacrum (Hepburn 1892; Sigmon 1974) and according to Sigmon (1974) also from the margin of the greater sciatic notch. The muscle descends through the greater sciatic foramen to an insertion at the anteromedial aspect of the tip of the greater femoral trochanter (Hepburn 1892; Beddard 1893; Sigmon 1974).
- Usual innervation: Branches of sacral plexus (Sigmon 1974).
- Notes: According to Beddard (1893) and Sigmon (1974) the muscle is always fused with the gluteus medius in hylobatids. However, in the three hylobatids in which we examined this region in detail (*H. lar* specimens HU HL1 and GWU HL1, *H. gabriellae* specimen VU HG2) the piriformis was not fused with the gluteus medius.

Quadratus femoris (3.9 g; Figs. 39, 40)

- Usual attachments: It originates from the lateral aspect of the ischial tuberosity near to the biceps femoris and to the semitendinosus (our dissections: *H. lar* specimens HU HL1 and GWU HL1, *H. gabriellae* specimen VU HG2). It is a small, thick, fleshy muscle (Beddard 1893; Sigmon 1974) separate from the adductor magnus and with a superior border that is closely related to the inferior gemellus (Hepburn 1892). It inserts onto the intertrochanteric crest (Hepburn 1892; Sigmon 1974) but it also extends to the posterior surface of the greater trochanter (Hepburn 1892).
- Usual innervation: Sacral nerves (Sigmon 1974).
- Notes: Hepburn (1892) stated that the **articularis genu** is present in all apes, but in the three hylobatids in which we examined this anatomical region in

detail (*H. lar* specimens HU HL1 and GWU HL1, *H. gabriellae* specimen VU HG2) the articularis genu is seemingly not present as a distinct muscle.

Rectus femoris (12.1 g; Figs. 40, 41, 32, 43, 47)

- Usual attachments: Hylobatids only have a caput rectum; the caput reflexum is absent (Hepburn 1892; Sigmon 1974). In the three hylobatids in which we examined this anatomical region in detail (*H. lar* specimens HU HL1 and GWU HL1, *H. gabriellae* specimen VU HG2) the caput rectum runs from the anterior inferior iliac spine to the patella.
- Usual innervation: Femoral nerve (our dissections: *H. lar* specimens HU HL1 and GWU HL1, *H. gabriellae* specimen VU HG2).

Vastus intermedius (27.2 g; Figs. 43, 47)

- Usual attachments: It runs from the ventral aspect of the femoral shaft (the proximal extension being narrow: Sigmon 1974) to the patella (our dissections: *H. lar* specimens HU HL1 and GWU HL1, *H. gabriellae* specimen VU HG2).
- Usual innervation: Femoral nerve (our dissections: *H. lar* specimens HU HL1 and GWU HL1, *H. gabriellae* specimen VU HG2).

Vastus lateralis (36.2 g; Figs. 43, 47)

- Usual attachments: It runs from the lateral aspect of the greater femoral trocanter and the distal two-thirds of the lateral femoral shaft in the region of the lateral lip of the linea aspera (the two origins being continuous: Beddard 1893; Sigmon 1974) to the patella (our dissections: *H. lar* specimens HU HL1 and GWU HL1, *H. gabriellae* specimen VU HG2).
- Usual innervation: Femoral nerve (our dissections: *H. lar* specimens HU HL1 and GWU HL1, *H. gabriellae* specimen VU HG2).

Vastus medialis (18.2 g; Figs. 40, 41, 42, 43, 47)

- Usual attachments: It originates from the dorsomedial femoral shaft in the region of the linea aspera (Beddard 1893; Sigmon 1974) more proximally than in other hominoids (Sigmon 1974); it also takes origin from the iliofemoral ligament (Beddard 1893; Sigmon 1974). It inserts onto the patella and the tibial tuberosity via the patellar tendon (Hepburn 1892; Beddard 1893) and together with the other vasti, the vastus medialis muscle contributes to the dorsolateral aspect of the knee capsule (Hepburn 1892).
- Usual innervation: Femoral nerve (our dissections: *H. lar* specimens HU HL1 and GWU HL1, *H. gabriellae* specimen VU HG2).

Sartorius (11.6 g; Figs. 40, 47, 49)

- Usual attachments: It originates from the anterior iliac border (Hepburn 1892; Beddard 1893), also referred to as 'the region of the anterior superior iliac spine' as well as from the lateral iliac border (Sigmon 1974). The ribbon-shaped muscle curses obliquely over the thigh dorsal to the medial femoral condyle (Hepburn 1892; Beddard 1893; Sigmon 1974). Insertion is onto the medial border of the tibial shaft (Hepburn 1892; Beddard 1893; Sigmon 1974) superficial to the insertion of the gracilis and of the semitendinosus (Hepburn 1892; Beddard 1893). In the three hylobatids in which we examined this region in detail (*H. lar* specimens HU HL1 and GWU HL1, *H. gabriellae* specimen VU HG2) the sartorius originates from the lateral iliac border of the anterior superior iliac spine, the insertion of the muscle being narrower than, and distal to, the insertion of the muscle in the other hominoids dissected by us.

- Usual innervation: Femoral nerve (Sigmon 1974).

Tensor fasciae latae (Figs. 47, 48)

- Usual attachments: It originates from the region of the anterior superior iliac spine (Kaplan 1958a; Sigmon 1974) and it may also originate from the gluteal fascia (Sigmon 1974). According to Hepburn (1892) and Sigmon (1974) the muscle is fused proximally with the gluteus maximus and according to the latter author it is fused laterally with the gluteus medius and the gluteus minimus (Sigmon 1974). However, in the three hylobatids in which we examined this region in detail (*H. lar* specimens HU HL1 and GWU HL1, *H. gabriellae* specimen VU HG2) we did not detect any fusion of the tensor fasciae latae with any of these muscles. The tensor fasciae latae inserts onto the iliotibial tract (Duvernoy 1855–1856; Hepburn 1892; Kaplan 1958a,b; Sigmon 1974).
- Usual innervation: The mediodorsal part of the tensor fasciae latae is innervated by the superior gluteal nerve, while the anterolateral part is innervated by the femoral nerve (Sigmon 1974).

Adductor brevis (2.1 g; Figs. 41, 42, 47, 49)

- Usual attachments: When present in hylobatids (Hepburn 1892) the muscle has a single origin from the body of the pubis (Hepburn 1892; Beddard 1893; Sigmon 1974), this origin lying between the origins of the gracilis and of the obturator externus (Hepburn 1892; Sigmon 1974). The adductor brevis is partly fused with the short head of the adductor magnus (Hepburn 1892; Sigmon 1974). Sigmon (1974) and Hepburn (1892) stated that in hylobatids the adductor brevis is usually at least partially fused with the adductor magnus, but in the three hylobatids in which we examined this anatomical region in detail (*H. lar* specimens HU HL1 and GWU HL1, *H. gabriellae* specimen VU HG2) there is seemingly no fusion with this latter muscle. The insertion of the adductor brevis lies between the insertions of the pectineus and of the superior part of the adductor magnus (Hepburn 1892).
- Usual innervation: Ventral division of obturator nerve (Sigmon 1974; Hamada 1985).

Adductor longus (3.6 g; Figs. 40, 49)

- Usual attachments: The origin is by a flat tendon from the anterior superior pubic ramus in the region of the pubic tubercle (Beddard 1893; Sigmon 1974). It inserts onto the middle of the medial lip of the linea aspera or posteromedial femoral shaft (Hepburn 1892; Beddard 1893; Appleton and Ghey 1929; Sigmon

1974), beside and ventral to the proximal half of the insertion of the short head of the adductor magnus (Beddard 1893; Sigmon 1974). The adductor longus is said to insert more proximally in *Hylobates* than in the great apes (Appleton and Ghey 1929).

- Usual innervation: Data are not available.

Adductor magnus (54.4 g; Figs. 40, 41, 49)

- Usual attachments: It has a continous origin from the anterior surface of the inferior pubic ramus lateral to the symphysis and from the inferior ischial ramus as far as the ischial tuberosity (Hepburn 1892; Beddard 1893; Sigmon 1974; Yirga 1987). According to Sigmon (1974) in 2/3 of *Hylobates* specimens the muscle divides into two parts distally; in the remaining 1/3 it remains joined until its insertion. The adductor magnus is described as fasciculated (Hepburn 1893; Beddard 1893) and may take origin from the medial border of the semitendinosus and from the long head of the biceps femoris according to Sigmon (1974). It inserts onto the medial lip of the linea aspera or the posteromedial surface of the femur and onto the adductor tubercle of the medial epicondyle (Hepburn 1892; Beddard 1893; Sigmon 1974). The attachment onto the linea aspera runs upwards to the inferior border of the insertion of the quadratus femoris (Hepburn 1892). In the three hylobatids in which we examined this anatomical region in detail (*H. lar* specimens HU HL1 and GWU HL1, *H. gabriellae* specimen VU HG2) the adductor magnus divides into two parts distally. The long head inserts onto the linea aspera but in these three specimens there is no adductor tubercle.
- Usual innervation: Short head by obturator nerve (Hepburn 1892; Sigmon 1974); long head by flexores femoris nerve (Sigmon 1974).
- Notes: The **adductor minimus** is usually absent in hylobatids (Sigmon 1974; our dissections), although according to Hepburn (1892) it may be 'present' but adherent to the other adductors.

Gracilis (10.9 g; Figs. 40, 47, 49, 50, 51)

- Usual attachments: It originates from the inferior pubic ramus near to the pubic symphysis onto the ischial ramus (Hepburn 1892; Beddard 1893; Sigmon 1974). This origin extends to the whole pubic body according to Hepburn (1892). It inserts onto the ventromedial surface of the tibia (Hepburn 1892; Sigmon 1974) and there is an aponeurotic expansion to the fascia of the leg (Hepburn 1892). This insertion lies between those of the sartorius and semitendinosus, being medial and distal to that of the sartorius and lateral and proximal to that of the semitendinosus (Hepburn 1892; Beddard 1893; Sigmon 1974).
- Usual innervation: Ventral branch of obturator nerve (Sigmon 1974).

Pectineus (3.4 g; Fig. 40)

- Usual attachments: It runs from the superior pubic ramus (Hepburn 1892; Sigmon 1974) to the dorsal surface of the femur just distal to the lesser trochanter

(Hepburn 1892; Beddard 1893; Sigmon 1974); there is a well-marked groove at its insertion in hylobatids and gorillas, but the groove is less conspicuous in *Pan* and *Pongo* (Appleton and Ghey 1929).

- Usual innervation: Femoral nerve and not by both the obturator and femoral nerves as is occasionally the case in *Pan* and *Pongo* (Sigmon 1974).
- Notes: To our knowledge, the **iliocapsularis** was not described as a distinct muscle in hylobatids.

Biceps femoris (caput longum + caput breve: 11.9 g; Figs. 38, 39, 42, 43, 50)
- Usual attachments: The long head of the biceps femoris originates from the ischial tuberosity in common with the semitendinosus (Beddard 1893; Prejzner-Morawska and Urbanowick 1971; Sigmon 1974; Hamada 1985; Kumakura 1989). According to Sigmon (1974) this origin also blends with the gluteus maximus, but in the three hylobatids in which we examined this region in detail (*H. lar* specimens HU HL1 and GWU HL1, *H. gabriellae* specimen VU HG2) there is no blending with this latter muscle. The short head of the biceps femoris originates from the dorsolateral femur in the region of the lateral lip of the linea aspera (Beddard 1893; Prejzner-Morawska and Urbanowick 1971; Sigmon 1974; Hamada 1985; Kumakura 1989), extending more distally in great apes than in *Hylobates* and *Homo* (Fick 1895a,b; Sigmon 1974). It also takes origin from the lateral intermusular septum (Beddard 1893; Prejzner-Morawska and Urbanowick 1971; Hamada 1985). The two heads of the muscle are fused in 7/9 *Hylobates* according to Gibbs (1999; based on Hepburn 1892; Prejzner-Morawska and Urbanowick 1971; Sigmon 1974; Kumakura 1989). The long head of the biceps femoris inserts onto the tibial head (Hepburn 1892; Prejzner-Morawska and Urbanowick 1971; Sigmon 1974), the tibial tuberosity or condyle (Sneath 1955; Kaplan 1958a,b; Sigmon 1974) and the fibular head and fascia of the leg (Beddard 1893; Prejzner-Morawska and Urbanowick 1971; Sigmon 1974; Hamada 1985; Kumakura 1989). This head may also insert onto the capsule of the knee joint according to Hepburn (1892) and Prejzner-Morawska and Urbanowick (1971). The short head inserts onto the fibular head and fascia of the leg (Hepburn 1892; Beddard 1893; Prejzner-Morawska and Urbanowick 1971; Sigmon 1974; Hamada 1985; Kumakura 1989) and onto the tibial tuberosity (Prejzner-Morawska and Urbanowick 1971; Sigmon 1974).
- Usual innervation: Caput longum by flexores femoris nerve as is usually the case in other apes (and not by the the tibial nerve, which is usually the case in modern humans); caput breve by common peroneal nerve as is usually the case in other apes and in modern humans (Sigmon 1974).

Semimembranosus (9.9 g; Figs. 40, 41, 42, 49)
- Usual attachments: Its origin is from the ischial tuberosity, inferior and lateral to the semitendinosus according to Sigmon (1974), but only inferior (not

lateral) to this latter muscle according to our dissections (*H. lar* specimens HU HL1 and GWU HL1, *H. gabriellae* specimen VU HG2). The insertion tendon is round and flat (Hepburn 1892), attaching onto the dorsal surface of the medial tibial condyle (Hepburn 1892; Beddard 1893; Sigmon 1974).

- Usual innervation: By the flexores femoris nerve, as is usually the case in other apes (not by the the tibial nerve, as is usually the case in modern humans).

Semitendinosus (13.0 g; Figs. 40, 41, 42, 49, 50)

- Usual attachments: It originates from the ischial tuberosity in common with the long head of the biceps femoris (Hepburn 1892; Sigmon 1974). An oblique tendinous intersection in the muscle belly of the semitendinosus is occasionally present (Macalister 1871; Hepburn 1892). The insertion is by a narrow, flat tendon (Beddard 1893) that according to Hepburn (1892) extends further distally than in *Homo* and *Gorilla*. The insertion is distal to that of gracilis (Hepburn 1892; Beddard 1893; Sigmon 1974; our *H. lar* specimen GWU HL1) although in some hylobatids it has been described as being more medial to that of gracilis (Sigmon 1974).
- Usual innervation: By the flexores femoris nerve, as is usually the case in the other apes (not by the the tibial nerve, as is usually the case in humans).

Extensor digitorum longus (4.5 g; Figs. 50, 52, 54)

- Usual attachments: It runs from the head and medial crest of the fibula and the intermuscular septum (Beddard 1893; Kaplan 1958a; Lewis 1966), as well as from the lateral tibial condyle (Beddard 1893; Lewis 1966) to the dorsal aponeurosis of digits 2–5 (Hepburn 1892; Beddard 1893; Lewis 1966). In one (*H. lar* specimen GWU HL1) of the three hylobatids in which we examined this anatomical region in detail (the other two being the *H. lar* specimen HU HL1 and the *H. gabriellae* specimen VU HG2) the tendons to each digit are not clearly separated from each other.
- Usual innervation: Data are not available.

Extensor hallucis longus (2.4 g; Figs. 52, 54)

- Usual attachments: It originates from the medial surface of the fibula (Beddard 1893; Lewis 1966) and also from the lateral tibial condyle according to Lewis (1966). Its tendon passes along the shaft of the metatarpal of the hallux (MI) (Hepburn 1892; Beddard 1893) and then inserts onto its terminal phalanx (Beddard 1893; Lewis 1966).
- Usual innervation: Data are not available.
- Notes: According to the literature review of Gibbs (1999), the **fibularis tertius** ('peroneus tertius') is present in 1/2 of *Hylobates* (based on Kaneff 1986; when it is present in primates, it usually has a fascial insertion onto mV), 30% of *Gorilla*, 5% of *Pan*, and 95% of modern humans, being absent in orangutans. However, in the three hylobatids in which we examined this anatomical region in detail

(*H. lar* specimens HU HL1 and GWU HL1, *H. gabriellae* specimen VU HG2) the fibularis tertius is missing.

Tibialis anterior (9.2 g; Figs. 50, 52, 54)
- Usual attachments: It originates from the lateral tibia (Beddard 1893; Lewis 1966) where it takes the form of two bellies whereas more distally they are less distinct (Deniker 1885; Hepburn 1892; Beddard 1893; Lewis 1966). The muscle inserts onto the plantar surface of the medial cuneiform (Hepburn 1892; Beddard 1893; Lewis 1966; this insertion being the larger of the two terminations) and onto the proximal end of MI (Hepburn 1892; Lewis 1966), although in juvenile *Hylobates* the secondary insertion has been given as a cartilaginous "prehallux" in the tarsometatarsal joint (Lewis 1966).
- Usual innervation: Data are not available.

Fibularis brevis (2.7 g; Figs. 45, 50, 51)
- Usual attachments: Runs from the anterior region of fibula (Lewis 1966) to the tuberosity at the base of MV (Hepburn 1892; Beddard 1893; Lewis 1966).
- Usual innervation: Data are not available.
- Synonymy: Peroneus brevis or peroneus brevis (Gibbs 1999).

Fibularis longus (4.2 g; Figs. 44, 45, 50, 51, 53, 55)
- Usual attachments: It originates from the fibular head and the proximal fibula shaft (Ruge 1878a; Beddard 1893; Lewis 1966) as well as from the lateral tibial condyle (Lewis 1966). Its tendon runs in a groove on the cuboid bone (Lewis 1966) and inserts onto the tuberosity of MI (Hepburn 1892; Beddard 1893; Lewis 1966). There is a fibrous attachment to MV according to Lewis (1966). The tendon of the fibularis longus often has a small sesamoid bone (Manners-Smith 1908).
- Usual innervation: Data are not available.
- Notes: In one (*H. gabriellae* specimen VU HG2) of the three hylobatids in which we examined this region in detail (the other two being the *H. lar* specimens HU HL1 and GWU HL1) the fibularis longus has an additional tendon inserting laterally onto the dorsal aponeurosis of digit 5 (Fig. 45)
- Synonymy: Peroneus longus or peronaeus longus (Gibbs 1999).

Gastrocnemius (Figs. 50, 51)
- Usual attachments: The muscle has two heads that originate from the medial and lateral femoral condyles and the capsule of the knee joint and it joins with the soleus (Hepburn 1892; Beddard 1893; Frey 1913) to insert onto the calcaneal tuberosity (Beddard 1893); the calf is relatively flat (Hepburn 1892). A single sesamoid bone is present in both heads of gastrocnemius in hylobatids; such bones are absent in the other apes and they are inconstant in modern humans (Hepburn 1892; Frey 1913; Gibbs 1999). Within the primates dissected by us, the tendo calcaneus is better defined and more substantial in

Hylobates and *Pongo* than it is in *Gorilla* and *Pan*, but it is not as well-defined as is usually the case in *Homo*.

- Usual innervation: Tibial nerve (our dissections: *H. lar* specimens HU HL1 and GWU HL1, *H. gabriellae* specimen VU HG2).

Plantaris (2.5 g; Figs. 44, 50, 51)

- Usual attachments: According to the literature review of Gibbs (1999) the plantaris is absent in hylobatids and gorillas, it is only occasionally present in *Pongo*, it is present in 3/5 of *Pan* and in 90–95% of modern humans. However, in the three hylobatids in which we examined this region in detail (*H. lar* specimens HU HL1 and GWU HL1, *H. gabriellae* specimen VU HG2) the plantaris is clearly present as a distinct muscle with an elongated belly and a long, thin tendon (Figs. 44, 50, 51) that runs from the lateral tibial condyle (superiorly to the origin of the lateral head of the gastrocnemius) to the medial side of the calcaneal tuberosity.
- Usual innervation: Data are not available.

Soleus (8.8 g; Figs. 44, 50, 51)

- Usual attachments: It runs from the head and superodorsal aspect of the fibular shaft (Hepburn 1892; Beddard 1893; Huxley 1864; Lewis 1962; the tibial origin is absent from *Hylobates* according to Hepburn 1892 and Lewis 1962) to the calcaneal tuberosity (Beddard 1893).
- Usual innervation: Tibial nerve (our dissections: *H. lar* specimens HU HL1 and GWU HL1, *H. gabriellae* specimen VU HG2).

Flexor digitorum longus (6.5 g; Figs. 44, 46, 50, 53, 55)

- Usual attachments: It originates from the posterior aspect of the tibial shaft (*H. lar* specimens HU HL1 and GWU HL1, *H. gabriellae* specimen VU HG2) where it is fuses to a lesser or greater degree with the flexor digitorum brevis (Boyer 1935). The distribution of the tendons to the digits is variable: digit 5 is said to be supplied in all hylobatids (Hepburn 1892; Beddard 1893; Keith 1894ab; Lewis 1962); digit 4 is also said to be supplied in all hylobatids (Hepburn 1892; Lewis 1962); digit 3 is said to be supplied in 1/3 hylobatids (Hepburn 1892); digit 2 is said to be supplied in 2/3 hylobatids (Lewis 1962) and in a hylobatid reported by Hepburn (1892) the muscle was described as being fused with the flexor hallucis longus. Within the three hylobatids in which we examined this muscle in detail, the *H. lar* specimen HU HL1 has tendons to digits 2, 4 and 5, the *H. lar* specimen GWU HL1 has tendons to digits 4 and 5 (Fig. 55) and the *H. gabriellae* specimen VU HG2 has tendons to digits 2, 3, 4 and 5 (Fig. 46).
- Usual innervation: Tibial nerve (our dissections: *H. lar* specimens HU HL1 and GWU HL1, *H. gabriellae* specimen VU HG2).
- Synonymy: Flexor digitorum medialis (Gibbs 1999).

Flexor hallucis longus (12.9 g; Figs. 44, 46, 50, 51, 52, 53, 55)

- Usual attachments: It runs from the interosseous membrane and posterior crural intermuscular septum (Beddard 1893) to the base of the distal phalanx of the hallux (Hepburn 1892; Beddard 1893; Keith 1894a,b; Lewis 1962). According to Hepburn (1892), Beddard (1893), Keith (1894a,b) and Lewis (1962) there are additional insertions onto digits 3 and 4. According to Keith (1894a,b) there is often also an insertion onto digit 2 and he also reported an insertion onto digit 5. Within the three hylobatids in which we examined this muscle in detail the *H. lar* specimen HU HL1 has tendons to digits 1, 3 and 4 (Fig. 53), while the *H. lar* specimen GWU HL1 and the *H. gabriellae* specimen VU HG2 have tendons to digits 1, 2, 3 and 4 (Figs. 46, 55).
- Usual innervation: Tibial nerve (our dissections: *H. lar* specimens HU HL1 and GWU HL1, *H. gabriellae* specimen VU HG2).
- Synonymy: Flexor digitorum lateralis (Gibbs 1999).

Popliteus (2.4 g; Figs. 44, 51)

- Usual attachments: It originates from the lateral femoral condylar (Hepburn 1892; Beddard 1893) and inserts onto the posterior surface of the tibia (Beddard 1893). A sesamoid bone in the tendon at the lateral tibial condyle is absent from all hylobatids (Kohlbrügge 1890–1892; Forster 1903; Van Westrienen 1907; Vallois 1914) except for a single gibbon reported by Pearson and Davin (1921).
- Usual innervation: Tibial nerve (Hepburn 1892).

Tibialis posterior (8.4 g; Figs. 51, 53)

- Usual attachments: It originates from the interosseous membrane and the adjoining sides of the tibia and fibula (Beddard 1893). Its medial part inserts onto the navicular bone (Hepburn 1892; Lewis 1964); the lateral part may insert onto MII, MIII or MIV, but not onto the cuboid (Lewis 1964). An insertion onto the plantar ligaments has been reported (Hepburn 1892; Beddard 1893) as has an insertion onto the sheath of the tendon of the fibularis longus (Hepburn 1892; Lewis 1964).
- Usual innervation: Tibial nerve (our dissections: *H. lar* specimens HU HL1 and GWU HL1, *H. gabriellae* specimen VU HG2).

Extensor digitorum brevis (1.7 g; Figs. 45, 54)

- Usual attachments: It originates from the calcaneus giving rise to three tendons each inserting into the dorsal aponeuroses of digits 2, 3 and 4, respectively (Ruge 1878a; Hepburn 1892; Beddard 1893; Lewis 1966). However, according to Lewis (1966) in 1/2 *Hylobates* there is an additional tendon for digit 5 that merges with the fibularis brevis. In the three hylobatids in which we examined this muscle in detail (calcaneous) the tendons insert onto the dorsal aponeurosis of digits 2, 3 and 4, but in VU HG2 there is an accessory fascicle running from the calcaneus to the lateral side of MII (Fig. 45).

- Usual innervation: Medial branch of peroneal nerve (our dissections: calcaneous).

Extensor hallucis brevis (0.6 g; Figs. 45, 54)
- Usual attachments: A separate muscle that originates from the extensor digitorum brevis according to Hepburn (1892), Beddard (1893) and Lewis (1966) and which according to Hepburn (1892) and Lewis (1966) inserts onto the base of the proximal phalanx of the hallux. However, in the three hylobatids in which we examined this region in detail (*H. lar* specimens HU HL1 and GWU HL1, *H. gabriellae* specimen VU HG2) the extensor hallucis brevis has a calcaneal origin.
- Usual innervation: Medial branch of peroneal nerve (our dissections: *H. lar* specimens HU HL1 and GWU HL1, *H. gabriellae* specimen VU HG2).

Abductor digiti minimi (Figs. 44, 46, 52)
- Usual attachments: To our knowledge, there are no detailed descriptions of this muscle in hylobatids. In the three hylobatids in which we examined this region in detail (*H. lar* specimens HU HL1 and GWU HL1, *H. gabriellae* specimen VU HG2) the muscle runs from the medial and lateral calcaneus and from the plantar aponeurosis to the base of MV, laterally to the proximal phalanx of digit 5. In HU HL1 there are two tendons of insertion: one begins to be a distinct near to the base of MV, while the other only begins to be distinct structure distally to the base of MV.
- Usual innervation: Lateral plantar nerve (our dissections: *H. lar* specimens HU HL1 and GWU HL1, *H. gabriellae* specimen VU HG2).
- Synonymy: Abductor digiti quinti (Gibbs 1999).

Abductor hallucis (2.6 g; Figs. 44, 46, 52, 53, 55)
- Usual attachments: It originates from the medial and plantar surfaces of the calcaneus (Brooks 1887; Hepburn 1892; Beddard 1893) and from the medial part of the plantar aponeurosis (Hepburn 1892; Beddard 1893), and inserts onto the base of the proximal phalanx of the hallux (Brooks 1887; Hepburn 1892; Beddard 1893). According to Brooks (1887) and Hepburn (1892), gibbons have a separate slip that goes to the base of MI, which has been named the '**abductor ossis metacarpi hallucis**'. It has a sesamoid bone in its tendon that amalgamates close to its insertion with the tibialis anterior (Brooks 1887). There is also some degree of fusion between the abductor hallucis and the flexor hallucis brevis (Bischoff 1870; Ruge 1878b; Brooks 1887; Beddard 1893).
- Usual innervation: Medial plantar nerve (Brooks 1887).
- Notes: The **abductor metatarsi quinti** (abductor ossis metacarpi digiti quinti or abductor os metatarsi digiti minimi *sensu* Gibbs 1999) has been described in great apes as being distinct from the abductor digiti minimi and it has been reported as a variant in modern humans (see, e.g., Gibbs 1999), but to our knowledge it has not been reported in hylobatids. In the three hylobatids in

which we examined this region in detail (*H. lar* specimens HU HL1 and GWU HL1, *H. gabriellae* specimen VU HG2) there was no évidence of an abductor metatarsi quinti.

Flexor digitorum brevis (0.7 g; Figs. 44, 52, 55)

- Usual attachments: It originates from the calcaneus, usually from its medial and ventral surfaces (Hepburn 1892; Beddard 1893; Sarmiento 1983); according to Hepburn (1892) there is also an origin from the plantar aspect of the flexor digitorum longus. The muscle is deep to the central plantar fascia (Hepburn 1892) and supplies tendons to digits 2 and 3 (Hartmann 1886; Hepburn 1892; Beddard 1893; Sarmiento 1983) and according to Hepburn (1892) and Sarmiento (1983) also providing a tendon to digit 4. When present, the tendon to digit 5 is derived from a deep head of the muscle and according to Hepburn (1892) and Sarmiento (1983) this head also supplies digits 3 and 4. Within the three hylobatids in which we examined this region in detail (*H. lar* specimens HU HL1 and GWU HL1, *H. gabriellae* specimen VU HG2), in HU HL1 the flexor digitorum brevis goes to digits 2, 3 and 4 (Fig. 52), in GWU HL1 a single tendon of insertion to digit 2 and in VU HG2 the muscle goes to digits 2, 3, 4, and 5 (Fig. 44).
- Usual innervation: Data are not available.
- Notes: To our knowledge the **quadratus plantae** (flexor accessorius *sensu* Gibbs 1999) has not been reported in hylobatids. In the three hylobatids in which we examined this region in detail (*H. lar* specimens HU HL1 and GWU HL1, *H. gabriellae* specimen VU HG2) this muscle is missing.

Lumbricales (lumbricalis I + lumbricalis II + lumbricalis III + lumbricalis IV: 1.4 g; Figs. 53, 55)

- Usual attachments: The lumbricales run from the tendons of the flexor digitorum longus and of the flexor hallucis longus to the medial side of the extensor aponeuroses of digits 2 (lumbricalis I), 3 (lumbricalis II), 4 (lumbricalis III) and 5 (lumbricalis IV); lumbricalis I always has a single head from the flexor digitorum longus tendon to digit 2, while according to Hepburn (1892) and Beddard (1893) lumbricalis II has a double origin,. Within the three hylobatids in which we we examined this region in detail (*H. lar* specimens HU HL1 and GWU HL1, *H. gabriellae* specimen VU HG2) in HU HL1 the lumbricalis II has a double origin from the flexor digitorum longus tendon to digit 2 and from the flexor hallucis longus tendon to digit 3, while the lumbricalis III is fused with the flexor hallucis longus tendon to digit 4 and with the lumbricalis IV, which is in turn fused with the flexor hallucis longus tendon to digit 4 and with the flexor digitorum longus tendon to digit 5 (Fig. 53).
- Usual innervation: Data are not available.

Adductor hallucis (caput transversum + caput obliquum: 3.1 g; Figs. 45, 46, 53, 54, 55)

- Usual attachments: The two heads of the adductor hallucis are always fused in hylobatids according to Brooks (1887). The oblique head originates from MII and MIII (Brooks 1887; Beddard 1893) with an additional origin from MIV and an origin from the interosseous fascia according to Brooks (1887). The transverse head originates from MII and MIII (Brooks 1887; Beddard 1893) although an origin from MIV is also present in hylobatids according to Gibbs (1999) and according to Beddard (1893) and Brooks (1887) there is also an origin from the third and fourth metatarsophalangeal joints, ligaments and interosseus fascia. The oblique head inserts onto the base of the proximal phalanx of the hallux (Brooks 1887) and onto MI (Brooks 1887; Beddard 1893), its tendon extending to the distal phalanx and inserting onto a sesamoid bone (Brooks 1887). The insertion of the transverse head is onto the base of the proximal phalanx of the hallux (Brooks 1887), and onto MI (Brooks 1887; Beddard 1893). The combined muscle extends to the distal phalanx, sesamoid bone and capsular ligaments according to Brooks (1887).
- Usual innervation: Deep branch of lateral plantar nerve (Brooks 1887; Hepburn 1892).
- Notes: In modern humans the '**transversalis pedis**' (not listed in Terminologia Anatomica 1998) corresponds to the caput transversum of the adductor hallucis.

Flexor digiti mimini brevis (0.9 g; Figs. 46, 52, 55)
- Usual attachments: To our knowledge there are no detailed descriptions of this muscle in hylobatids. In the three hylobatids in which we examined this region in detail (*H. lar* specimens HU HL1 and GWU HL1, *H. gabriellae* specimen VU HG2) the muscle originates from the plantar tarsometatarsal ligament extending to the tuberosity of MV and inserts onto the base and the medial and lateral sides of the proximal phalanx of digit 5.
- Usual innervation: Data are not available.
- Synonymy: Flexor digiti quinti brevis or flexor digiti minimi (Gibbs 1999).

Flexor hallucis brevis (g; Figs. 44, 46, 52, 53, 55)
- Usual attachments: This muscle has two heads that are almost equal in size (Brooks 1887; Hepburn 1892; Beddard 1893) and according to Brooks (1887) in gibbons it is fused with the abductor hallucis. The entire muscle inserts onto the proximal phalanx of the hallux, while there are additional insertions onto MI (Brooks 1887; Beddard 1893). In the three hylobatids in which we examined this region in detail (*H. lar* specimens HU HL1 and GWU HL1, *H. gabriellae* specimen VU HG2) only the lateral head fuses with the abductor hallucis; this head runs from the plantar cuneonavicular ligament to the medial side of MI, while the medial head runs from the cuneiform I (os cuneiforme mediale) to the central and medial sides of MI.
- Usual innervation: Medial plantar nerve (Brooks 1887).

Opponens digiti minimi (Fig. 55)
- Usual attachments: To our knowledge, there are no detailed descriptions of this muscle in hylobatids. In the three hylobatids in which we examined this region in detail (*H. lar* specimens HU HL1 and GWU HL1, *H. gabriellae* specimen VU HG2) the muscle runs from the medial side of the tuberosity at the base of MV (its origin being closely related with the origin of the fourth plantar interossei) to the base of the proximal phalanx (via the medial sesamoid bone) of digit 5.
- Usual innervation: Data are not available.
- Notes: The opponens digiti minimi is usually considered to be a deep fascicle of the flexor digiti minimi in orangutans and modern humans (for instance, it was not listed in Terminologia Anatomica 1998), but it has been described as a distinct muscle in *Pan* and *Gorilla* (see, e.g., Gibbs 1999). The **opponens hallucis** has been described as a variant in hylobatids (Bischoff 1870), but in the three hylobatids in which we examined this region in detail (*H. lar* specimens HU HL1 and GWU HL1, *H. gabriellae* specimen VU HG2) there is no evidence of this structure.
- Synonymy: Opponens digiti quinti (Gibbs 1999).

Interossei dorsales (I + II + III + IV: 1.5 g; Figs. 45, 52, 54)
- Usual attachments: According to Brooks (1887) each dorsal interosseous has two heads of origin, with the exception of the first dorsal interosseous, which has a single head of origin from the medial side of MII; also according to him, the insertion of each muscle is extended to the distal phalanx via the extensor aponeurosis. In the three hylobatids in which we examined this region in detail (*H. lar* specimens HU HL1 and GWU HL1, *H. gabriellae* specimen VU HG2) there are four dorsal interossei. In GWU HL1 and VU HG2 the first dorsal interosseous has a single head of origin from the medial side of MII (Figs. 45, 54), but in HU HL1 this muscle originates from the lateral side and the medial side of MI and of MII, respectively (Fig. 52). In all three specimens the second dorsal interosseous originates from the lateral side of MII and the medial side of MIII, the third dorsal interosseous originates from the lateral side of MIII and the medial side of MIV and the fourth dorsal interosseous originates from the lateral side of MIV and medial side of MV. In all three specimens the first dorsal interosseous inserts onto the medial side of the base of the proximal phalanx of digit 2, the second dorsal interosseous inserts onto the lateral side of the base of the proximal phalanx of digit 3, the third dorsal interosseous inserts onto the lateral side of the base of the proximal phalanx of digit 3, and the fourth dorsal interosseous inserts onto the lateral side of the base of the proximal phalanx of digit 4. Therefore, the reference pedal digit is digit 3 as as is usually the case in great apes, rather than digit 2 as is usually the case in *Homo* (e.g., Gibbs 1999).
- Usual innervation: Data are not available.

Interossei plantares

- Usual attachments: To our knowledge, there are no detailed descriptions of these muscle in hylobatids. In the three hylobatids in which we examined this region in detail (*H. lar* specimens HU HL1 and GWU HL1, *H. gabriellae* specimen VU HG2) the first plantar interosseous runs from the medial side of MII to the medial side of the base of the proximal phalanx of digit 2, the second plantar interosseous runs from the medial side of MIV to the medial side of the base of the proximal phalanx of digit IV, and the third plantar interosseous runs from the medial side of MV to the medial side of the base of the proximal phalanx of digit V. Therefore, the reference digit is digit 3, as as is usually the case in great apes, rather than digit 2 as is usually the case in *Homo* (e.g., Gibbs 1999).
- Usual innervation: Data are not available.

Appendix I
Literature Including Information about the Muscles of Hylobatids*

Aiello L, Dean C (1990) *An introduction to human evolutionary anatomy*. San Diego: Academic Press.

Andrews P, Groves CP (1976) Gibbons and brachiation. In *Gibbon and Siamang, Vol. 4* (ed. Rumbaugh DM), pp. 167–218. Basel: Karger.

Appleton AB, Ghey PHR (1929) An example of the cervico-costo-humeral muscle of Gruber. *J Anat* 63, 434–436.

Ashton EH, Oxnard CE (1963) The musculature of the primate shoulder. *Trans Zool Soc Lond* 29, 553–650.

Ashton EH and Oxnard CE (1964) Functional adaptations of the primate shoulder girdle. *Proc Zool Soc Lond* 142, 49–66.

Aversi-Ferreira TA, Diogo R, Potau JM, Bello G, Pastor JF, Aziz MA (2010) Comparative anatomical study of the forearm extensor muscles of *Cebus libidinosus* (Rylands et al. 2000; Primates, Cebidae), modern humans, and other primates, with comments on primate evolution, phylogeny and manipulatory behavior. *Anat Rec* 293, 2056–2070.

Aziz MA, Dunlap SS (1986) The human extensor digitorum profundus muscle with comments on the evolution of the primate hand. *Primates* 27, 293–319.

Barnard WS (1875) Observations on the membral musculation of *Simia satyrus* (Orang) and the comparative myology of man and the apes. *Proc Amer Assoc Adv Sci* 24, 112–144.

Beddard FE (1893) Contributions to the anatomy of the anthropoid apes. *Trans Zool Soc Lond* 13, 177–218.

Bischoff TLW (1870) Beitrage zur Anatomie des *Hylobates leuciscus* and zueiner vergleichenden Anatomie der Muskeln der Affen und des Menschen. *Abh Bayer Akad Wiss Miinchen Math Phys Kl* 10, 197–297.

Blake ML (1976) *The quantitative myology of the hind limb of Primates with special reference to their locomotor adaptations*. Unpublished PhD Thesis, University of Cambridge, Cambridge.

Bojsen-Møller F (1978) Extensor carpi radialis longus muscle and the evolution of the first intermetacarpal ligament. *Am J Phys Anthropol* 48, 177–184.

Bolk L (1902) Beiträge zur Affenanatomie, III, Der Plexus cervico-brachialis der Primaten. *Petrus Campter* 1, 371–566.

---

*List not exhaustive

Boyer EL (1935) The musculature of the inferior extremity of the orang-utan, *Simia satyrus*. *Am J Anat* 56, 192–256.

Broca P (1869) L'ordre des primates—parallele anatomique de l'homme et des singes. *Bull Soc Anthropol Paris* 4, 228–401.

Brooks HSJ (1886a) On the morphology of the intrinsic muscles of the little finger, with some observations on the ulnar head of the short flexor of the thumb. *J Anat Physiol* 20, 644–661.

Brooks HSJ (1886b) Variations in the nerve supply of the flexor brevis pollicis muscle. *J Anat Physiol* 20, 641–644.

Brooks HSJ (1887) On the short muscles of the pollex and hallux of the anthropoid apes, with special reference to the opponens hallucis. *J Anat Physiol* 22, 78–95.

Brown B (1983) An evaluation of primate caudal musculature in the identification of the ischiofemoralis muscle. *Am J Phys Antropol* 60, 177–178.

Burrows AM (2008) The facial expression musculature in primates and its evolutionary significance. *Bioessays* 30, 212–225.

Burrows AM, Diogo R, Waller BM, Jonar CJ, Liebal K (2011). Morphology of facial expression musculature in a monogamous ape: evaluating the relative influences of ecological and phylogenetic factors in hylobatids. *Anat Rec* 294, 645–663.

Cave AJE (1979) The mammalian temporo-pterygoid ligament. *J Zool Lond* 188, 517–532.

Champneys F (1872) The muscles and nerves of a Chimpanzee (*Troglodytes Niger*) and a *Cynocephalus Anubis*. *J Anat Physiol* 6, 176–211.

Chapman HC (1900) Observations upon the anatomy of *Hylobates leuciscus* and *Chiromys Madagascariensis*. *Proc Acad Nat Sci Phila* 52, 414–423

Clegg M (2001) *The comparative anatomy and evolution of the human vocal tract.* Unpublished PhD Thesis, University of London, London.

Day MH, Napier J (1963) The functional significance of the deep head of flexor pollicis brevis in primates. *Folia Primatol* 1, 122–134.

Dean MC (1984) Comparative myology of the hominoid cranial base, I, the muscular relationships and bony attachments of the digastric muscle. *Folia Primatol* 43, 234–48.

Dean MC (1985) Comparative myology of the hominoid cranial base, II, the muscles of the prevertebral and upper pharyngeal region. *Folia Primatol* 44, 40–51.

Deniker J (1885) Recherches anatomiques et embryologiques sur les singes anthropoides, foetus de gorille et de gibbon. *Arch Zool Exp Génerale* 3, 1–265.

Diogo R, Abdala V (2010). Muscles of vertebrates—comparative anatomy, evolution, homologies and development. Oxford: Taylor and Francis.

Diogo R, Wood BA (2008). Comparative anatomy, phylogeny and evolution of the head and neck musculature of hominids: a new insight. *Am J Phys Anthropol, Suppl* 46, 90.

Diogo R, Wood BA (2009) Comparative anatomy and evolution of the pectoral and forelimb musculature of primates: a new insight. *Am J Phys Anthropol, Meeting Suppl* 48, 119.

Diogo R, Wood BA (2011a). Comparative anatomy and phylogeny of primate muscles and human evolution. Oxford: Taylor and Francis.

Diogo R, Wood BA (2011b). Soft-tissue anatomy of the primates: phylogenetic analyses based on the muscles of the head, neck, pectoral region and upper limb, with notes on the evolution of these muscles. *J Anat in press.*

Diogo R, Abdala V, Lonergan N, Wood BA (2008) From fish to modern humans—comparative anatomy, homologies and evolution of the head and neck musculature. *J Anat* 213, 391–424.

Diogo R, Abdala V, Aziz MA, Lonergan N, Wood BA (2009a) From fish to modern humans—comparative anatomy, homologies and evolution of the pectoral and forelimb musculature. *J Anat* 214, 694–716.

Diogo R, Wood BA, Aziz MA, Burrows A (2009b) On the origin, homologies and evolution of primate facial muscles, with a particular focus on hominoids and a suggested unifying nomenclature for the facial muscles of the Mammalia. *J Anat* 215, 300–319.

Donisch E (1973) A comparative study of the back muscles of gibbon and man. In *Gibbon and Siamang, Vol. 2* (ed. Rumbaugh DM), pp. 99–120. Basel: Karger.

DuBrul EL (1958) *Evolution of the speech apparatus.* Springfield: Thomas.

Duckworth WLH (1904) *Studies from the Anthropological Laboratory, the Anatomy School,* Cambridge. London: C. J. Clay and Sons.

Duckworth WLH (1912) On some points in the anatomy of the plica vocalis. *J Anat Physiol* 47, 80–115.

Duckworth WLH (1915) *Morphology and anthropology (2nd ed.).* Cambridge: Cambridge University Press.

Dunlap SS, Thorington RW, Aziz MA (1985) Forelimb anatomy of New World monkeys: myology and the interpretation of primitive anthropoid models. *Am J Phys Anthropol* 68, 499–517.

Duvernoy M (1855-1856) Des caracteres anatomiques de grands singes pseudoanthropomorphes anthropomorphes. *Arch Mus Natl Hist Nat Paris* 8, 1–248.

Dylevsky I (1967) Contribution to the ontogenesis of the flexor digitorum superficialis and the flexor digitorum profundus in man. *Folia Morphol (Praha)* 15, 330–335.

Dwight T (1895) Notes on the dissection and brain of the chimpanzee 'Gumbo'. *Mem Boston Soc Nat Hist* 5, 31–51.

Edgeworth FH (1935) *The cranial muscles of vertebrates.* Cambridge: Cambridge University Press.

Elftman HO (1932) The evolution of the pelvic floor of primates. *Am J Anat* 51, 307–346.

Fabre P-H, Rodrigues A, Douzery EJP (2009). Patterns of macroevolution among Primates inferred from a supermatrix of mitochondrial and nuclear DNA. *Mol Phyl Evol* 53, 808–825.

Fick R (1895a) Vergleichend-anatomische Studien an einem erwachsenen Orang-utang. *Arch Anat Physiol Anat Abt* 1895, 1–100.

Fick R (1895b) Beobachtungen an einem zweiten envachsenen Orang-Utang und einem Schimpansen. *Arch Anat Physiol Anat Abt* 1895, 289–318.

Fitzwilliams DCL (1910) The short muscles of the hand of the agile gibbon (*Hylobatis agilis*), with comments on the morphological position and function of the short muscles of the hand of man. *Proc R Soc Edinb* 30, 201–218.

Fleagle JG (1999) *Primate Adaptation and Evolution (2nd Ed.).* San Diego: Academic Press.

Fleagle JG, Stern JT, Jungers WL, Susman RL, Vangor AK, Wells JP (1981) Climbing: A biomechanical link with brachiation and bipedalism. *Symp zool Soc Lond* 48, 359–373.

Forster A (1903) Die Insertion des Musculus semimembranosus. *Arch Anat Physiol Anat Abt* 1953, 257–320.

Forster A (1917) Die mm. contrahentes und interossei manus in der Säugetierreihe und beim Menschen. *Arch Anat Physiol Anat Abt* 1916, 101–378.

Forster A (1933) Contribution à l'evolution du pouce chez *Hylobates leuciscus. Arch Anat Histiol Embryol* 16, 215–230.

Frey H (1913) Der Musculus triceps surae in der Primatenreihe. *Morph Jahrb* 47, 1–192.

Giacomini C (1897) 'Plica semilunaris' et larynx chez les singes anthropomorphes. *Arch Ital Biol* 28, 98–119.

Gibbs S (1999) *Comparative soft tissue morphology of the extant Hominoidea, including Man.* Unpublished PhD Thesis, The University of Liverpool, Liverpool.

Gibbs S, Collard M, Wood BA (2000) Soft-tissue characters in higher primate phylogenetics. *Proc Natl Acad Sci US* 97, 11130–11132.

Gibbs S, Collard M, Wood BA (2002) Soft-tissue anatomy of the extant hominoids: a review and phylogenetic analysis. *J Anat* 200, 3–49.

Grönroos H (1903) Die musculi biceps brachii und latissimocondyloideus bei der affengattung *Hylobates* im vergleich mit den ensprechenden gebilden der anthropoiden und des menschen. *Abh Kön Preuss Akad Wiss Berlin* 1903, 1–102.

Groves CP (1986) Systematics of the great apes. In *Comparative Primate Biology: Systematics, Evolution and Anatomy, Vol. 1* (eds. Swindler DR, Erwin J), pp. 187–217. New York: A.R. Liss.

Groves CP (1995) *Revised character descriptions for Hominoidea.* Typescript, 9 pp.

Hamada Y (1985) Primate hip and thigh muscles: comparative anatomy and dry weights. In *Primate Morphophysiology, Locomotor Analyses and Human Bipedalism* (ed. Kondo S), pp. 131–152. Tokyo: University of Tokyo Press.

Hänel H (1932) Über die Gesichtsmuskulatur der katarrhinen Affen. *Gegenbaur Morph Jahrb* 71, 1–76.

Harrison DFN (1995) *The anatomy and physiology of the mammalian larynx.* Cambridge: University Press.

Hartmann R (1886) *Anthropoid apes.* London: Keegan.

Hepburn D (1892) The comparative anatomy of the muscles and nerves of the superior and inferior extremities of the anthropoid apes: I—Myology of the superior extremity. *J Anat Physiol* 26, 149–186.

Hepburn D (1896) A revised description of the dorsal interosseous muscles of the human hand, with suggestions for a new nomenclature of the palmar interosseous muscles and some observations on the corresponding muscles in the anthropoid apes. *Trans R Soc Edin* 38, 557–565.

Herring SW, Herring SE (1974) The superficial masseter and gape in mammals. *Am Nat* 108, 561–576.

Hill WCO, Harrison-Matthews L (1949) The male external genitalia of the gorilla, with remarks on the os penis of other Hominoidea. *Proc Zool Soc Lond* 119, 363–378.

Hill WCO, Kanagasuntheram R (1959) The male reproductive organs in certain gibbons (*Hylobatidae*). *Am J Phys Anthropol* 17, 227–241.

Hofër W (1892) Vergleichend-anatomische Studien uber die Nerven des Armes und der Hand bei den Affen und dem Menschen. *Munchener Med Abhandl* 30, 1–106.

Howell AB (1936a) Phylogeny of the distal musculature of the pectoral appendage. *J Morphol* 60, 287–315.

Howell AB (1936b) The phylogenetic arrangement of the muscular system. *Anat Rec* 66, 295–316.

Howell AB, Straus WL (1932) The brachial flexor muscles in primates. *Proc US Natl Mus* 80, 1–31.

Huber E (1930a) Evolution of facial musculature and cutaneous field of trigeminus—Part I. *Q Rev Biol* 5, 133–188.

Huber E (1930b) Evolution of facial musculature and cutaneous field of trigeminus—Part II. *Q Rev Biol* 5, 389–437.

Huber E (1931) *Evolution of facial musculature and expression.* Baltimore: The Johns Hopkins University Press.

Huxley TH (1864) The structure and classification of the Mammalia. *Med Times Gazette* 1864, 398–468.

Huxley TH (1871) *The anatomy of vertebrated animals.* London; J. and A. Churchill.

Imparati E (1895-1896) Contribuzione alia miologia delle regione antero-laterale del torace, costale, e della spalla, nelle Seimmie. *Eiv Ital Sci Nat Siena* 15, 118–121,129–132,145–148;16, 7–9,17–24.

Jouffroy FK (1971) Musculature des membres. In *Traité de Zoologie, XVI: 3 (Mammifères)* (ed. Grassé PP), pp. 1–475. Paris: Masson et Cie.

Jouffroy FK, Lessertisseur J (1959) Reflexions sur les muscles contracteurs des doigts et des orteils (contrahentes digitorum) chez les primates. *Ann Sci Nat Zool, Ser 12*, 1, 211–235.

Jouffroy FK, Lessertisseur J (1960) Les spécialisations anatomiques de la main chez les singes à progression suspendue. *Mammalia* 24, 93–151.

Jouffroy FK, Saban R (1971) Musculature peaucière. In *Traité de Zoologie, XVI: 3 (Mammifères)* (ed. Grassé PP), pp. 477–611. Paris: Masson et Cie.

Juraniec J (1972) The aortic and esophageral hiatus in the diaphragm of primates. *Folia Morphol* 31, 197–207.

Juraniec J, Szostakiewicz-Sawicka H (1968) The central tendon of the diaphragm in primates. *Folia Morphol* 27, 183–194.

Jungers WL, Stern JT (1980) Telemetered electromyography of forelimb muscle chains in gibbons (*Hylobates lar*). *Science* 208, 617–619.

Jungers WL, Stern JT (1981) Preliminary electromyographical analysis of brachiation in gibbon and spider monkey. *Int J Primatol* 2, 19–33.

Jungers WL, Stern JT (1984) Kinesiological aspects of brachiation in lar gibbons. In *The Lesser Apes: Evolutionary and Behavioral Biology* (eds. Preuschoft H, Chiver DJ, Brockelman WY, Creel N), pp. 119–134. Edinburgh: Edinburgh University Press.

Kanagasuntheram R (1952–1954) Observations on the anatomy of the hoolock gibbon. *Ceylon J Sci Sect G* 5, 11–64+69–122.

Kaneff A (1959) Über die evolution des m. abductor pollicis longus und m. extensor pollicis brevis. *Mateil morphol Inst Bulg Akad Wiss* 3, 175–196.

Kaneff A (1968) Zur differenzierung des m. abductor pollicis biventer beim Menschen. *Gegenbaurs morphol Jahrb* 112, 289–303.

Kaneff A (1969) Umbildung der dorsalen Daumenmuskeln beim Menschen. *Verh Anat Ges* 63, 625–636.

Kaneff A (1979) Évolution morphologique des musculi extensores digitorum et abductor pollicis longus chez l'Homme. I. Introduction, méthodologie, M. extensor digitorum. *Gegenbaurs Morphol Jahrb* 125, 818–873.

Kaneff A (1980a) Évolution morphologique des musculi extensores digitorum et abductor pollicis longus chez l'Homme. II. Évolution morphologique des m. extensor digiti minimi, abductor pollicis longus, extensor pollicis brevis et extensor pollicis longus chez l'homme. *Gegenbaurs Morphol Jahrb* 126, 594–630.

Kaneff A (1980b) Évolution morphologique des musculi extensores digitorum et abductor pollicis longus chez l'Homme. III. Évolution morphologique du m. extensor indicis chez l'homme, conclusion générale sur l'évolution morphologique des musculi extensores digitorum et abductor pollicis longus chez l'homme. *Gegenbaurs Morphol Jahrb* 126, 774–815.

Kaneff A (1986) Die Aufrichtung des Menschen und die mor-phologisches Evolution der Musculi extensores digitorum pedis unter dem Gesichtpunkt der evolutiven Myologie, Teil I. *Morph Jahrb* 132, 375–419.

Kaneff A, Cihak R (1970) Modifications in the musculus extensor digitorum lateralis in phylogenesis and in human ontogenesis. *Acta Anat Basel* 77, 583–604.

Kaplan EB (1958a) The iliotibial tract—clinical and morphological significance. *J Bone Jt Surg* 40A, 817–832.

Kaplan EB (1958b) Comparative anatomy of the extensor digitorum longus in relation to the knee joint. *Anat Rec* 131, 129–149.

Keith A (1891) Anatomical notes on Malay apes. *J Strts Brit Roy Assoc Soc* 23, 77–94.

Keith A (1894a) *The myology of the Catarrhini: a study in evolution.* Unpublished PhD thesis, University of Alberdeen, Alberdeen.

Keith A (1894b) Notes on a theory to account for the various arrangements of the flexor profundus digitorum in the hand and foot of primates. *J Anat Physiol* 28, 335–339.

Keith A (1896) A variation that occurs in the manubrium stemi of higher primates. *J Anat Phys* 30, 275–279.

Kikuchi Y (2010a) Comparative analysis of muscle architecture in primate arm and forearm. *Anat Histol Embryol* 39, 93–106.

Kikuchi Y (2010b) Quantitative analysis of variation in muscle internal parameters in crab-eating macaques (*Macaca fascicularis*). *Anthropol Sci* 118, 9–21.

Kleinschmidt A (1950) Zur anatomie des kehlkopfs der Anthropoiden. *Anat Anz* 97, 367–372.

Kohlbrügge JHF (1890–1892) Versuch einer Anatomie des Genus *Hylobates*. In *Zoologische Ergebnisse Einer Reise in Niederländisch Ost-Indien* (ed. Weber M), pp. 211–354 (Vol. 1), 138–208 (Vol. 2). Leiden: Verlag von EJ Brill.

Kohlbrügge JHF (1896) Der larynx und die stimmbildung der Quadrumana. *Natuurk T Ned Ind* 55, 157–175.

Kohlbrügge JHF (1897) Muskeln und Periphere Nerven der Primaten, mit besonderer Berücksichtigung ihrer Anomalien. *Verh K Akad Wet Amsterdam Sec 2* 5, 1–246.

Koizumi M, Sakai T (1995) The nerve supply to coracobrachialis in apes. *J Anat* 186, 395–403.

Kumakura H (1989) Functional analysis of the biceps femoris muscle during locomotor behavior in some primates. *Am J Phys Anthropol* 79, 379–391.

Laitman JT (1977) *The Ontogenetic and phylogenetic development of the upper respiratory system and basicranium in man.* Unpublished PhD thesis, Yale University, New Haven.

Lander KF (1918) The pectoralis minor: a morphological study. *J Anat* 52, 292–318.

Landsmeer JM(1984) The human hand in phylogenetic perspective. *Bull Hosp Jt Dis Orthop Inst* 44, 276–287.

Landsmeer JM (1986) A comparison of fingers and hand in *Varanus*, opossum and primates. *Acta Morphol Neerl Scand* 24, 193–221.

Landsmeer JM (1987) The hand and hominisation. *Acta Morphol Neerl Scand* 25, 83–93.

Lewis OJ (1962) The comparative morphology of M. flexor accessorius and the associated long flexor tendons. *J Anat* 96, 321–333.

Lewis OJ (1964) The evolution of the long flexor muscles of the leg and foot. In *International Review of General and Experimental Zoology* (eds. Felts WJL, Harrison RJ), pp. 165–185. New York: Academic Press.

Lewis OJ (1965) The evolution of the Mm. interossei in the primate hand. *Anat Rec* 153, 275–287.

Lewis OJ (1966) The phylogeny of the cruropedal extensor musculature with special reference to the primates. *J Anat* 100, 865–880.

Lewis OJ (1989) *Functional morphology of the evolving hand and foot.* Oxford: Clarendon Press.

Lorenz R (1974) On the thumb of the Hylobatidae. In *Gibbon and Siamang, Vol. 3* (ed. Rumbaugh DM), pp. 157–175. Basel: Karger.

Loth E (1912) Beiträge zur Anthropologie der Negerweichteile (Muskelsystem). *Stud Forsch Menschen-u Völkerkunde Stuttgart* 9, 1–254.

Loth E (1931) *Anthropologie des parties molles (muscles, intestins, vaisseaux, nerfs peripheriques).* Paris: Mianowski-Masson et Cie.

Lunn HF (1948) The comparative anatomy of the inguinal ligament. *Anat Phys* 82, 58–67.

Lunn HF (1949) Observations on the mammalian inguinal region. *Proc Zool Soc* 118, 345–355.

Macalister A (1871) On some points in the myology of the chimpanzee and others of the primates. *Ann Mag Nat Hist* 7, 341–351.

Maier W (2008) Epitensoric position of the chorda tympani in Anthropoidea: a new synapomorphic character, with remarks on the fissura glaseri in Primates. In *Mammalian Evolutionary Morphology: a Tribute to Frederick S. Szalay* (eds. Sargis EJ, Dagosto M), pp. 339–352. Dordrecht: Springer.

Mangini U (1960) Flexor pollicis longus muscle: its morphology and clinical significance. *J Bone Jt Surg* 42A, 467–559.

Manners-Smith T (1908) A study of the cuboid and os perineum in the primate foot. *J Anat Phys* 42, 397–414.

McMurrich JP (1903a) The phylogeny of the forearm flexors. *Amer J Anat* 2, 177–209.

McMurrich JP (1903b) The phylogeny of the palmar musculature. *Amer J Anat* 2, 463–500.

Michilsens F, Vereecke EE, D'Août K, Aerts P (2009) Functional anatomy of the gibbon forelimb: adaptations to a brachiating lifestyle. *J Anat* 215, 335–354.

Mijsberg WA (1923) Über den Bau des Urogenitalapparates bei den männlichen Primaten. *Verh K Akad Wet Amsterdam* 23, 1–92.

Miller RA (1932) Evolution of the pectoral girdle and forelimb in the primates. *Amer J Phys Anthropol* 17, 1–56.

Miller RA (1934) Comparative studies upon the morphology and distribution of the brachial plexus. *Am J Anat* 54, 143–175.

Miller RA (1945) The ischial callosities of primates. *Am J Anat* 76, 67–87.

Miller RA (1947) The inguinal canal of primates. *Am J Anat* 90, 117–142.

Morton DJ (1922) Evolution of the human foot, part 1. *Am J Phys Anthropol* 5, 305–336.

Mysberg WA (1917) Über die Verbinderungen zwischen dem Sitzbeine und der Wirbelsäule bei den Säugetieren. *Anat Hefte* 54, 641–668.

Negus VE (1949) *The comparative anatomy and physiology of the larynx*. New York: Hafner Publishing Company.

Owen R (1868) *The Anatomy of Vertebrates, Vol. 3: Mammals*. London: Longmans, Green and Co.

Parsons FG (1898a) The muscles of mammals, with special relation to human myology, Lecture 1, The skin muscles and muscles of the head and neck. *J Anat Physiol* 32:428–450.

Parsons FG (1898b) The muscles of mammals, with special relation to human myology: a course of lectures delivered at the Royal College of Surgeons of England—lecture II, the muscles of the shoulder and forelimb. *J Anat Physiol* 32, 721–752.

Payne RC (2001) *Musculoskeletal adaptations for climbing in hominoids and their role as exaptations for the acquisition of bipedalism*. Unpublished PhD thesis, The University of Liverpool, Liverpool.

Pearson K, Davin AG (1921) On the sesamoids of the knee joint. *Biometrika* 13, 133–175, 350–400.

Plattner F (1923) Über die ventral-innervierte und die genuine rückenmuskulatur bei drei anthropomorphen (*Gorilla gina, Hylobates* und *Troglodytes niger*). *Morphol Jb* 52, 241–280.

Potau JM, Artells R, Bello G, Muñoz C, Monzó M, Pastor JF, de Paz F, Barbosa M, Diogo R, Wood B (in press) Expression of myosin heavy chain isoforms in the supraspinatus muscle of different primate species. *Int J Prim.*

Prejzner-Morawska A, Urbanowicz M (1971) The biceps femoris muscle in lemurs and monkeys. *Folia Morphol* 30, 9465–482.

Primrose A (1899) The anatomy of the orang-outang (*Simia satyrus*), an account of some of its external characteristics, and the myology of the extremities. *Trans Royal Can Inst* 6, 507–594.

Ranke K (1897) Muskel-und Nervenvariationen der dorsalen elemente des Plexus ischiadicus der Primaten. *Arch Anthropol* 24, 117–144.

Rauwerdink GP (1993) Muscle fibre and tendon lengths in primate extremities. In *Hands of Primates* (eds. Preuschoft H, Chivers DJ), pp. 207–223. New York: Springer-Verlag.

Rex H (1887) Ein Beitrag zur Kenntnis der Muskulatur der Mundspalte der Affen. *Morphol Jahrb* 12: 275–286.

Robinson JT, Freedman L, Sigmon BA (1972) Some aspects of pongid and hominid bipedality. *J Hum Evol* 1, 361–369.

Ruge G (1878a) Untersuchung uber die Extensorengruppe aus Unterschenkel und Füsse der Säugethiere. *Morphol Jahrb* 4, 592–643.

Ruge G (1878b) Zur vergleichenden Anatomie der tiefen Muskeln in der Fusssohle. *Morphol Jahrb* 4, 644–659.

Ruge G (1885) Über die Gesichtsmuskulatur der halbaffen. *Gegen Morph Jahrb* 11, 243–315.

Ruge G (1887a) *Untersuchungen uber die Gesichtsmuskeln der Primaten*. Leipzig: W. Engelmann.

Ruge G (1887b) Die vom Facialis innervirten Muskeln des Halses, Nackens und des Schädels einen jungen *Gorilla*. *Gegenb Morph Jahrb* 12, 459–529.

Ruge G (1890–1891) Anatomisches über den Rumpf der Hylobatiden—ein Beitrag zur Bestimmung der Stellung dieses genus im System. In *Zoologische Ergebnisse Einer Reise in Niederländisch Ost-Indien, Vol. 1* (ed. Weber M), pp. 366–460. Leiden: Verlag von EJ Brill.

Ruge G (1897) *Über das peripherische gebiet des nervus* facialis *boi wirbelthieren*. Leipzig: Festschr f Gegenbaur.

Ruge G (1911) Gesichtsmuskulatur und Nervus facialis der Gattung *Hylobates*. *Morph Jahrb* 44: 129–177.

Saban R (1968) Musculature de la tête. In Traité de Zoologie, XVI: 3 (Mammifères) (ed. Grassé PP), pp. 229–472. Paris: Masson et Cie.

Sarmiento EE (1983) The significance of the heel process in anthropoids. *Int J Primatol* 4, 127–152.

Schreiber HV (1934) Zur morphologie der primatenhand—rontnenolonische untersuchungen an der handwuriel dei affen. *Anat Anz* 78, 369–429.

Schreiber HV (1936) Die extrembewegungen der schimpansenhand, 2, mitteilung zu—zur morphologie der primatehand. *Morph Jahrb* 77, 22–60.

Schück AC (1913a) Beiträge zur Myologie der Primaten, I—der m. lat. dorsi und der m. latissimo-tricipitalis. *Morphol Jahrb* 45, 267–294.

Schück AC (1913b) Beiträge zur Myologie der Primaten, II—1 die gruppe sterno-cleido-mastoideus, trapezius, omo-cervicalis, 2 die gruppe levator scapulae, rhomboides, serratus anticus. *Morphol Jahrb* 46, 355–418.

Schultz AH (1936) Characters common to higher primates and characters specific for man. *Q Rev Biol* 11, 259–283, 425–455.

Schultz AH (1973) The skeleton of the Hylobatidae and other observations on their morphology. In *Gibbon and Siamang, Vol. 2* (ed. Rumbaugh DM), pp. 1–53. Basel: Karger.

Seiler R (1970) Differences in the facial musculature of the nasal and upper-lip region in catarrhine primates and man. *Z Morphol Anthropol* 62, 267–275.

Seiler R (1971a) A comparison between the facial muscles of Catarrhini with long and short muzzles. *Proc 3rd Int Congr Primat Zürich 1970, vol l, Basel: Karger*, 157–162.

Seiler R (1971b) Facial musculature and its influence on the facial bones of catarrhine Primates, I. *Morphol Jahrb* 116, 122–142.

Seiler R (1971c) Facial musculature and its influence on the facial bones of catarrhine Primates, II. *Morphol Jahrb* 116, 147–185.

Seiler R (1971d) Facial musculature and its influence on the facial bones of catarrhine Primates, III. *Morphol Jahrb* 116, 347–376.

Seiler R (1971e) Facial musculature and its influence on the facial bones of catarrhine Primates, IV. *Morphol Jahrb* 116, 456–481.

Seiler R (1976) Die Gesichtsmuskeln. In *Primatologia, Handbuch der Primatenkunde, Bd. 4, Lieferung 6* (eds. Hofer H, Schultz AH, Starck D), pp. 1–252. Basel: Karger.

Seiler R (1977) Morphological and functional differentiation of muscles—studies on the m. frontalis, auricularis superior and auricularis anterior of primates including man. *Verh Anat Ges* 71, 1385–1388.

Seiler R (1979a) Criteria of the homology and phylogeny of facial muscles in primates including man, I, Prosimia and Platyrrhina. *Morphol Jahrb* 125, 191–217.

Seiler R (1979b) Criteria of the homology and phylogeny of facial muscles in primates including man, II, Catarrhina. *Morphol Jahrb* 125, 298–323.

Seiler R (1980) Ontogenesis of facial muscles in primates. *Morphol Jahrb* 126, 841–864.

Shoshani J, Groves CP, Simons EL, Gunnell GF (1996) Primate phylogeny: morphological vs molecular results. *Mol Phylogenet Evol* 5, 102–154.

Shrewsbury MM, Marzke MM, Linscheid RL, Reece SP (2003) Comparative morphology of the pollical distal phalanx. *Am J Phys Anthropol* 121, 30–47.

Shrivastava RK (1978) *Anatomie comparée du muscle deltoide et son innervation dans la série des mammifères.* Unpublished Phd thesis, Université de Paris, Paris.

Sigmon BA (1974) A functional analysis of pongid hip and thigh musculature. *J Hum Evol* 3, 161–185.

Smith WC (1923) The levator ani muscle; its structure in man, and its comparative relationships. *Anat Rec* 26, 175–204.

Sneath RS (1955) The insertion of the biceps femoris. *J Anat* 89, 550–553.

Sonntag CF (1924) *The morphology and evolution of the apes and man.* London: John Bale Sons and Danielsson, Ltd.

Stern JT (1972) Anatomical and functional specializations of the human gluteus maximus. *Anta J Phys Anthropol* 36, 315–340.

Starck D, Schneider R (1960) Respirationsorgane. In *Primatologia III/2* (eds. Hofer H, Schultz AH, Starck D), pp. 423–587. Basel: Karger.

Stern JT, Larson SG (2001) Telemetered electromyography of the supinators and pronators of the forearm in gibbons and chimpanzees: implications for the fundamental positional adaptation of hominoids. *Am J Phys Anthopol* 115, 253–268.

Stern JT, Wells JP, Jungers WL, Vangor AK, Fleagle JG (1980a) An electromyographic study of the pectoralis major in atelines and *Hylobates* with special reference to the evolution of a pars clavicularis. *Am J Phys Anthropol* 52, 13–25.

Stern JT, Wells JP, Jungers WL, Vangor AK (1980b) An electromyographic study of serratus anterior in atelines and *Alouatta*: implications for hominoid evolution. *Am J Phys Anthropol* 52, 323–334.

Stewart TD (1936) The musculature of the anthropoids, I, neck and trunk. *Am J Phys Anthropol* 21, 141–204.

Stout K (2000) *Grip types and associated morphology of the hylobatid thumb and index finger*. Unpublished MA thesis, Arizona State University, Phoenix.

Straus WL (1941a) The phylogeny of the human forearm extensors. *Hum Biol* 13, 23–50.

Straus WL (1941b) The phylogeny of the human forearm extensors (concluded). *Hum Biol* 13, 203–238.

Straus WL (1942a) The homologies of the forearm flexors: urodeles, lizards, mammals. *Am J Anat* 70, 281–316.

Straus WL (1942b) Rudimentary digits in primates. *Q Rev Biol* 17, 228–243.

Susman RL (1994) Fossil evidence for early hominid tool use. *Science* 265, 1570–1573

Susman RL (1998) Hand function and tool behavior in early hominids. *J Hum Evol* 35, 23–46.

Susman RL, Jungers WL, Stern JT (1982) The functional morphology of the accessory interosseous muscle in the gibbon hand: determination of locomotor and manipulatory compromises. *J Anat* 134, 111–120.

Susman RL, Nyati L, Jassal MS (1999) Observations on the pollical palmar interosseus muscle (of Henle). *Anat Rec* 254, 159–165.

Tappen NC (1955) Relative weights of some functionally important muscles of the thigh, hip and leg in a gibbon and in man. *Am J Phys Anthropol* 13, 415–420.

Testut L (1883) Le long fléchisseur propre du pouce chez l'homme et les singes. *Bull Soc Zool Fr* 8, 164–185.

Testut L (1884) *Les anomalies musculaires chez l'homme expliquèes par l'anatomie comparée et leur importance en anthropologie*. Paris: Masson.

Thompson P (1901) On the arrangement of the fasciae of the pelvis and their relationship to the levator ani. *J Anat Phys* 35, 127–141.

Tocheri MW, Orr CM, Jacofsky MC, Marzke MW (2008) The evolutionary history of the hominin hand since the last common ancestor of *Pan* and *Homo*. *J Anat* 212, 544–562.

Tschachmachtschjan H (1912) Über die Pectoral- und Abdominal-musculatur und über die Scalenus-Gruppe bei Primataten. *Morph Jb* 44, 297–370.

Tuttle RH (1967) Knuckle-walking and the evolution of hominoid hands. *Am J Phys Anthrop* 26, 171–206.

Tuttle RH (1969) Quantitative and functional studies on the hands of the Anthropoidea, I, the Hominoidea. *J Morphol* 128, 309–363.

Tuttle RH (1972a) Relative mass of cheiridial muscles in catarrhine primates. In *The Functional and Evolutionary Biology of Primates* (ed. Tuttle RH), pp. 262–291. Chicago: Aldine-Atherdon.

Tuttle RH (1972b) Functional and evolutionary biology of hylobatid hands and feet. In *Gibbon and Siamang, Vol. 1* (ed. Rumbaugh DM), pp. 136–206. Basel: S. Karger.

Tuttle RH, Basmajian JV (1976) Electromyography of pongid shoulder muscles and hominoid evolution I—retractors of the humerus and rotators of the scapula. *Yearbook Phys Anthropol* 20, 491–497.

Tuttle RH, Basmajian JV (1978a) Electromyography of pongid shoulder muscles II—deltoid, rhomboid and "rotator cuff". *Am J Phys Anthropol* 49, 47–56.

Tuttle RH, Basmajian JV (1978b) Electromyography of pongid shoulder muscles III—quadrupedal positional behavior. *Am J Phys Anthropol* 49, 57–70.

Vallois H (1914) *Étude anatomique de l'articulation du genou ches les Primates*. Montpellier: L'Abeille.

Van den Broek AJP (1909) Ein doppelseitiger M.sternalis und ein M.pectoralis quartus bei *Hylobates syndactylus*. *Anat Anz* 35, 591–596.

Van den Broek AJP (1914) Studien zur Morphologie des Primatenbeckens. *Morphol Jahrb* 49, 1–118.

Van Horn RN (1972) Structural adaptations to climbing in the gibbon hand. *Am Anthropol New Ser* 74, 326–334.

Van Westrienen A (1907) Das Kniegelenk der Primaten, mit besonderer Berücksichtigung der Anthropoiden. *Petrus Camper* 4, 1–60.

Walmsley R (1937) The sheath of the rectus abdominis. *J Anat Phys* 77, 404–414.

Verhulst J (2003) *Developmental dynamics in humans and other primates: discovering evolutionary principles through comparative morphology*. Ghent: Adonis Press.

Vrolik W (1841) *Recherches d' anatornie comparé, sur le chimpanzé*. Amsterdam: Johannes Miller.

Wall CE, Larson SG, Stern JT (1994) EMG of the digastric muscle in gibbon and orangutan: functional consequences of the loss of the anterior digastric in orangutans. *Am J Phys Anthropol* 94, 549–567.

Whitehead PF (1993) Aspects of the anthropoid wrist and hand. In *Postcranial Adaptation in Nonhuman Primates* (ed. Gebo DL), pp 96–120. DeKalb: Northern Illinois University Press.

Winckler G (1950) Contribution a l'étude des muscles larges de la paroi abdominale, Étude d'anatomie comparée. *Arch Anat Histol Embryol* 33, 157–228.

Wood Jones F (1920) *The principles of anatomy as seen in the hand*. London: J. and A Churchill.

Yirga S (1987) Interrelation between ischium, thigh extending muscles and locomotion in some primates. *Primates* 28, 79–86.

Ziegler AC (1964) Brachiating adaptations of chimpanzee upper limb musculature. *Am J Phys Anthropol* 22, 15–32.

Zihlman AL, Brunker L (1979) Hominid bipedalism: then and now. *Yearb Phys Anthropol* 22, 132–162.

Appendix II
Literature Cited, not Including Information about the Muscles of Hylobatids

Arnold C, Matthews LJ, Nunn CL (2010) The 10kTrees Website: A New Online Resource for Primate Phylogeny. *Evol Anthropol* 19, 114–118.

Diogo R (2004a) *Morphological evolution, aptations, homoplasies, constraints, and evolutionary trends: catfishes as a case study on general phylogeny and macroevolution*. Enfield: Science Publishers.

Diogo R (2004b) Muscles versus bones: catfishes as a case study for an analysis on the contribution of myological and osteological structures in phylogenetic reconstructions. *Anim Biol* 54, 373–391.

Diogo R (2007) *On the origin and evolution of higher-clades: osteology, myology, phylogeny and macroevolution of bony fishes and the rise of tetrapods*. Enfield: Science Publishers.

Diogo R (2008) Comparative anatomy, homologies and evolution of the mandibular, hyoid and hypobranchial muscles of bony fish and tetrapods: a new insight. *Anim Biol* 58, 123–172.

Diogo R (2009) The head musculature of the Philippine colugo (Dermoptera: *Cynocephalus volans*), with a comparison to tree-shrews, primates and other mammals. *J Morphol* 270, 14–51.

Diogo R, Abdala V (2007) Comparative anatomy, homologies and evolution of the pectoral muscles of bony fish and tetrapods: a new insight. *J Morphol* 268, 504–517.

Diogo R, Potau JM, Pastor JF, de Paz FJ, Ferrero EM, Bello G, Barbosa M, Wood B (2010) Photographic and Descriptive Musculoskeletal Atlas of Gorilla. Oxford: Taylor and Francis.

Goodman M (1999) The natural history of the primates. *Amer J Hum Genet* 64, 31–39.

Goodman M, Braunitzer G, Stangl A, Schrank B (1983) Evidence on human origins from haemoglobins of African apes. *Nature* 303, 546–548.

Groves CP (2001) *Primate Taxonomy*. Washington, DC: Smithsonian Institution Press.

Groves CP (2005) Order Primates, Order Monotremata, (and select other orders). In *Mammal Species of the World, 3rd ed* (ed. Wilson DE, Reeder DM), pp. 178–181. Baltimore: Johns Hopkins University Press.

House EL (1953) A myology of the pharyngeal region of the albino rat. *Anat Rec* 116, 363–381.

Kelemen G (1948) The anatomical basis of phonation in the chimpanzee. *J Morphol* 82, 229–256.

Kelemen G (1969) Anatomy of the larynx and the anatomical basis of vocal performance. In *The Chimpanzee—Vol 1—Anatomy, behaviour and diseases of chimpanzees* (ed. Bourne GH), pp. 35–186. Basel: Karger.

Lightoller GHS (1925) Facial muscles—the modiolus and muscles surrounding the rima oris with some remarks about the panniculus adiposus. *J Anat Physiol* 60, 1–85.

Lightoller GS (1928) The facial muscles of three orang utans and two cercopithecidae. *J Anat* 63, 19–81.

Lightoller GS (1934) The facial musculature of some lesser primates and a *Tupaia*. *Proc Zool Soc Lond* 1934, 259–309.

Lightoller GS (1939) V. Probable homologues. A study of the comparative anatomy of the mandibular and hyoid arches and their musculature—Part I. Comparative myology. *Trans Zool Soc Lond* 24, 349–382.

Lightoller GS (1940a) The comparative morphology of the platysma: a comparative study of the sphincter colli profundus and the trachelo-platysma. *J Anat* 74, 390–396.

Lightoller GS (1940b) The comparative morphology of the m. caninus. *J Anat* 74, 397–402.

Lightoller GS (1942) Matrices of the facialis musculature: homologization of the musculature in monotremes with that of marsupials and placentals. *J Anat* 76, 258–269.

Napier JR, Napier PH (1985) *The Natural History of the Primates*. Cambridge: MIT Press.

Netter FH (2006) *Atlas of human anatomy (4th ed.)*. Philadelphia: Saunders.

Nowak RM (1999) Walker's Primates of the World. Baltimore: Johns Hopkins University Press.

Owen R (1830–1831) On the anatomy of the orangutan (*Simia satyrus*, L.). *Proc Zool Soc Lond* 1, 4–5, 9–10, 28–29, 66–72.

Raven HC (1950) Regional anatomy of the *Gorilla*. In *The anatomy of the* Gorilla (ed. Gregory WK), pp. 15–188.

Terminologia Anatomica (1998) *Federative Committee on Anatomical Terminology*. Stuttgart: Georg Thieme. New York: Columbia University Press.

Index

About the Authors

Rui Diogo is an Assistant Professor at the Howard University College of Medicine and a Resource Faculty at the Center for the Advanced Study of Hominid Paleobiology of George Washington University (US). He is the author or co-author of numerous publications, and the co-editor of the books *Catfishes* and *Gonorynchiformes and ostariophysan interrelationships—a comprehensive review*. He is the sole author or first author of the books *Morphological evolution, aptations, homoplasies, constraints and evolutionary trends—catfishes as a case study on general phylogeny and macroevolution, The origin of higher clades—osteology, myology, phylogeny and evolution of bony fishes and the rise of tetrapods Muscles of vertebrates—comparative anatomy, evolution, homologies and development, Photographic and descriptive musculoskeletal atlas of Gorilla—with notes on the attachments, variations, innervation, synonymy and weight of the muscles*, and *Comparative anatomy and phylogeny of primate muscles and human evolution*.

Josep Potau is University Professor at the Department of Anatomy and Embryology of the University of Barcelona (Spain) and is the director of the University's Center for the Study of Comparative and Evolutionary Anatomy. His current research focus on the analysis of functional and anatomical adaptations associated with the evolution of different types of locomotion and of the upper limb musculature within primates. He has published several papers and book chapters on functional and comparative anatomy.

Juan Pastor is University Professor at the Department of Anatomy of the University of Valladolid (Spain) and is the director of the University's Anatomical Museum, which houses the largest comparative osteological collection in Spain. He published several papers on comparative anatomy and anthropology.

Félix de Paz is University Professor at the Department of Anatomy of the University of Valladolid and is a member of the Royal Academy of Medicine and Surgery of Valladolid (Spain). He has published several papers on comparative anatomy and anthropology.

Mercedes Barbosa is University Professor at the Department of Anatomy of the University of Valladolid (Spain), and is a member of the Anatomical Society of Spain. She published several papers on physical anthropology.

Eva Ferrero is a biologist who graduated from the University of León (Spain) and obtained her PhD at the University of Valladolid (Spain). She is undertaking research at the Department of Anatomy and the Anatomical Museum of the University of Valladolid (Spain) on the comparative anatomy of primates and other mammals.

Gaëlle Bello is a biologist who graduated from the University of La Coruña (Spain). She undertook a Master in Primatology at the University of Barcelona (Spain), and is now undertaking a PhD at the University of Barcelona that focuses on the evolution of the scapula within Primates and its adaptations to different types of locomotion.

Anne M. Burrows is a biological anthropologist and an Associate Professor in the Department of Physical Therapy at Duquesne University and a Research Associate Professor in the Department of Anthropology at the University of Pittsburgh. Her current research focuses on the evolution of facial musculature in primates and the evolution of feeding specializations in primates. She has published numerous papers and book chapters on these subjects and has recently co-edited *The Evolution of Exudativory in Primates*.

M. Ashraf Aziz is University Professor of Anatomy at the Department of Anatomy of Howard University College of Medicine (USA). His research focuses on the comparative gross and developmental morphology of modern human aneuploidy syndromes, the evolution of the muscles supplied by the trigeminal nerve and of the arm and hand muscles in non-human and human primates and the value of human cadaver dissections/prosections in the age of digital information systems. He has published numerous papers in international journals, including *Teratology*, *American Journal of Physical Anthropology*, *Journal of Anatomy*, *The Anatomical Record* and *Primates*.

Julia Arias-Martorell is a biologist now undertaking a PhD at the University of Barcelona (Spain) that focuses on functional morphology and variabilty of the forelimbs of the hominoids related to the diverse locomotor repertoires of the members of this clade, and on the enhancement of tridimensional techniques to study evolution.

Bernard Wood is University Professor of Human Origins and directs the Center for the Advanced Study of Hominid Paleobiology at George Washington University (USA). His edited publications include *Food Acquisition and Processing in Primates* and *Major Topics in Primate and Human Evolution* and he is the author of *The Evolution of Early Man, Human Evolution, Koobi Fora Research Project—Hominid Cranial Remains (Vol. 4), Human Evolution—A Very Short Introduction*. He is the editor of the *Wiley-Blackwell Encyclopedia of Human Evolution*.

Color Plate Section

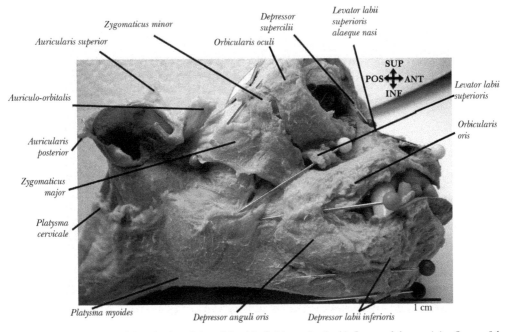

Fig. 1 *Hylobates lar* (HU HL1, adult male): lateral view of the right facial muscles. In this figure and the remaining figures of the atlas, the names of the muscles are in italics, and SUP, INF, ANT, POS, MED, LAT, VEN, DOR, PRO and DIS refer to superior, inferior, anterior, posterior, medial, lateral, ventral, dorsal, proximal and distal, respectively (in the sense the terms are applied to pronograde tetrapods: see Methodology and Material).

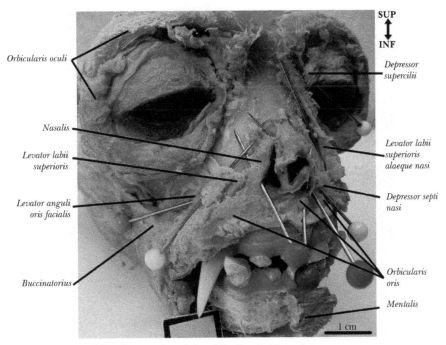

Fig. 2 *Hylobates lar* (HU HL1, adult male): anterolateral view of the deep facial muscles.

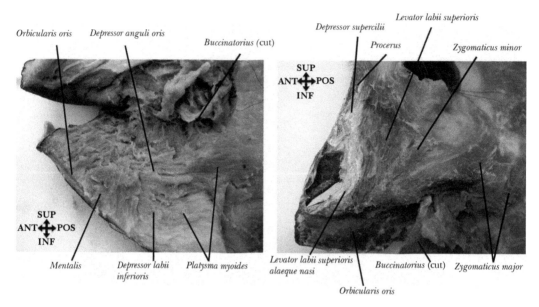

Fig. 3 *Hylobates syndactylus* (DU HS1, adult male): mesial view of the right facial mask (on the left; N.B., the green color delineates the boundary of the depressor labii inferioris); *Hylobates muelleri* (DU HM1, adult male): medial view of the right facial mask (on the right; N.B., the green and blue colors delineate the boundaries of the levator labii superioris alaeque nasi and of the levator labii superioris, respectively).

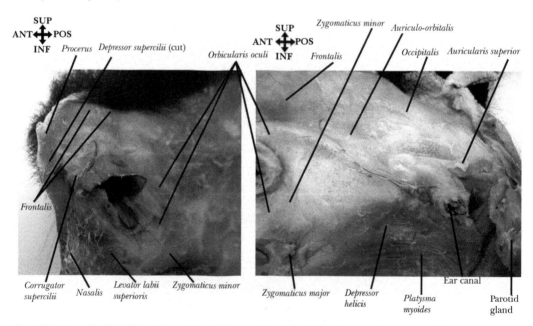

Fig. 4 *Hylobates muelleri* (DU HM1, adult male): medial views of the right facial masks (N.B., the green and blue colors delineate the boundaries of the corrugator supercilii and of the depressor supercilii, respectively).

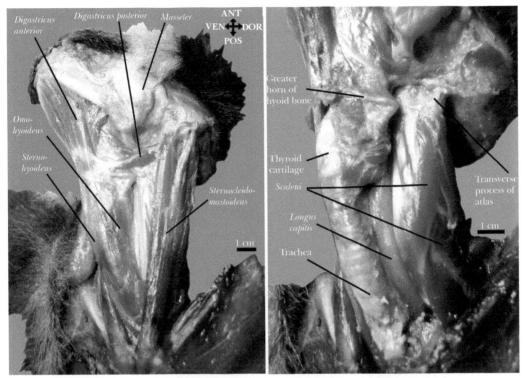

Fig. 5 *Hylobates gabriellae* (VU HG2, adult male): lateral views of the left head and neck muscles (N.B., on the right picture, the sternocleidomastoideus and the infrahyoid muscles were removed).

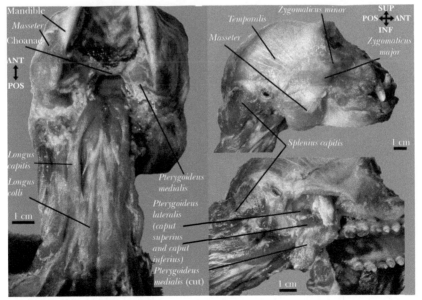

Fig. 6 *Hylobates gabriellae* (VU HG2, adult male): ventral view of the deep neck muscles after removal of pharinx, larinx and thachea (on the left); lateral and ventrolateral views of the left (horizontal flip done with photoshop) superficial (top, right) and deep (bottom, right) masticatory muscles.

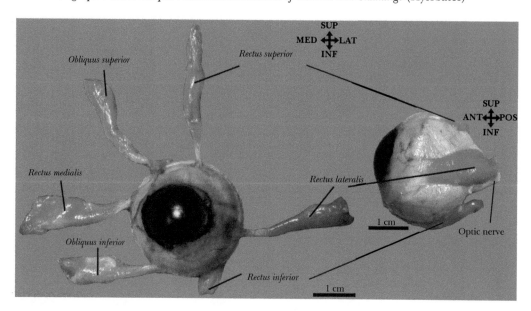

Fig. 7 *Hylobates gabriellae* (VU HG2, adult male): frontal (on the right) and lateral (on the left) views of the left eye ball and extra-ocular muscles.

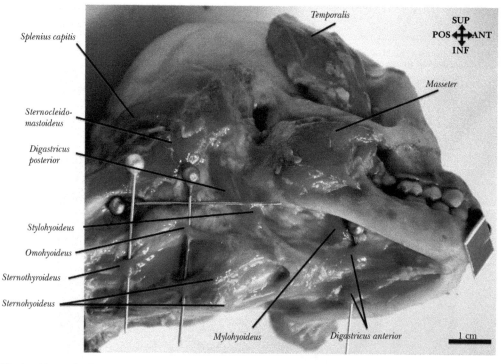

Fig. 8 *Hylobates gabriellae* (VU HG1, infant male): ventrolateral view of the right head and neck muscles, after removal of facial muscles.

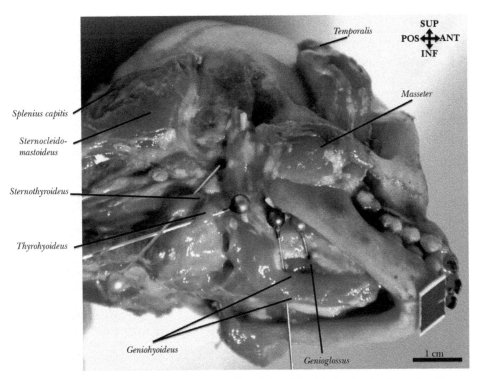

Fig. 9 *Hylobates gabriellae* (VU HG1, infant male): ventrolateral view of the right head and neck muscles, after removal of facial muscles and of digastricus posterior, digastricus anterior, mylohyoideus, stylohyoideus, omohyoideus and sternohyoideus.

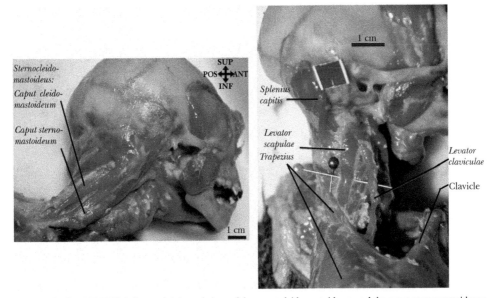

Fig. 10 *Hylobates gabriellae* (VU HG1, infant male): lateral views of the caput cleidomastoideum and the caput sternomastoideum of the right sternocleidomastoideus (on the left) and of the right trapezius, levator scapulae and levator claviculae (on the right).

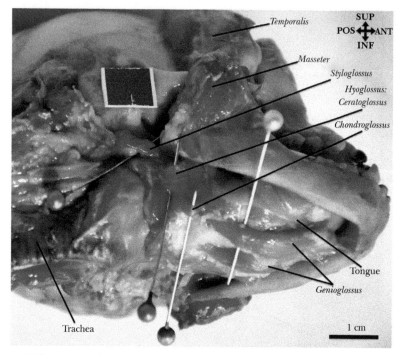

Fig. 11 *Hylobates gabriellae* (VU HG1, infant male): ventrolateral view of the right extrinsic tongue muscles; the masseter and temporalis are also shown.

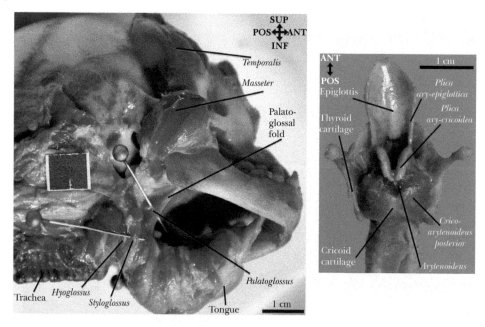

Fig. 12 *Hylobates gabriellae* (VU HG1, infant male): ventrolateral view of the right palatoglossal fold, including a few muscular fibers that seem to correspond to the fibers of the palatoglossus of humans (on the left; N.B., the tongue was moved inferiorly and rotated); *Hylobates gabriellae* (VU HG2, adult male): dorsal view of the laryngeal musculature (on the right).

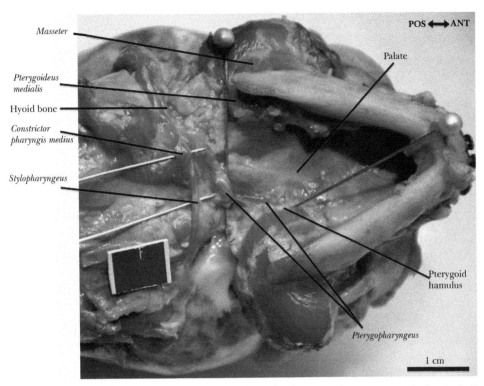

Fig. 13 *Hylobates gabriellae* (VU HG1, infant male): ventral view showing a muscular structure running longitudinally from the pterygoid hamulus to the region of the medial constrictor, forming the pterygo-pharyngeus, which is distinct from the superior constrictor due to the markedly longitudinal (antero-posterior) direction of its fibers; the hyoid bone and larynx were moved to show the pterygopharyngeus.

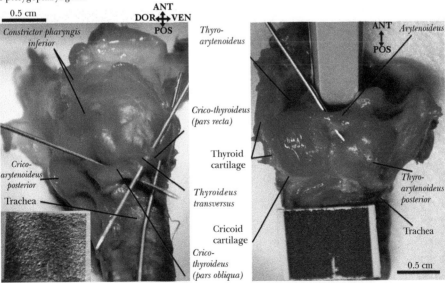

Fig. 14 *Hylobates gabriellae* (VU HG1, infant male): lateral view of the right laryngeal musculature (on the left); dorsal view of the laryngeal musculature (on the right).

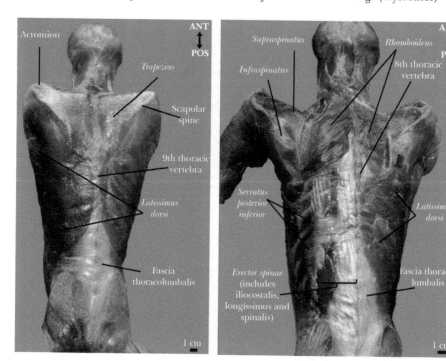

Fig. 15 *Hylobates gabriellae* (VU HG2, adult male): dorsal views of the superficial (on the left) and deeper (on the right) muscles of the back. Note that the erector spinae includes the muscles iliocostalis, longissimus and spinalis.

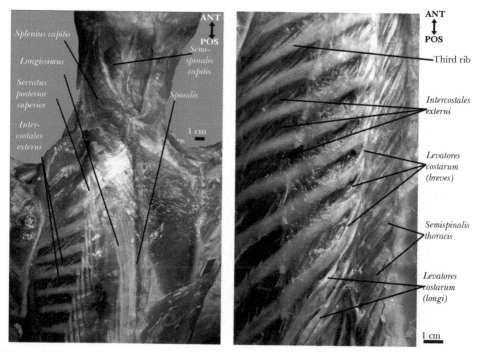

Fig. 16 *Hylobates gabriellae* (VU HG2, adult male): dorsal views of the deep muscles of the neck and back.

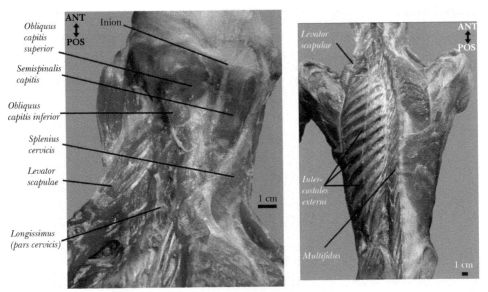

Fig. 17 *Hylobates gabriellae* (VU HG2, adult male): dorsal views of deep neck muscles (on the left) and of the deep back muscles after removal of serratus posterior inferior, serratus posterior superior, and erector spinae (on the right).

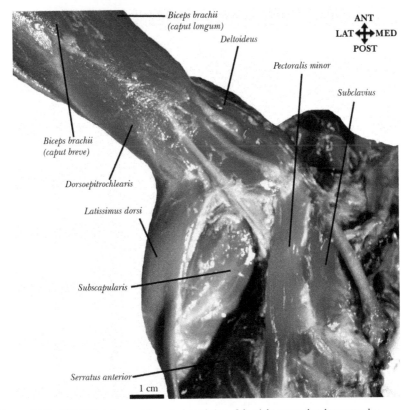

Fig. 18 *Hylobates gabriellae* (VU HG1, infant male): ventrolateral view of the right pectoral and arm muscles.

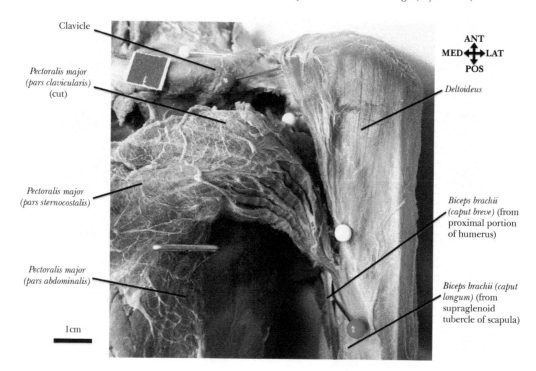

Fig. 19 *Hylobates lar* (HU HL1, adult male): ventral view of the left pectoral and arm musculature.

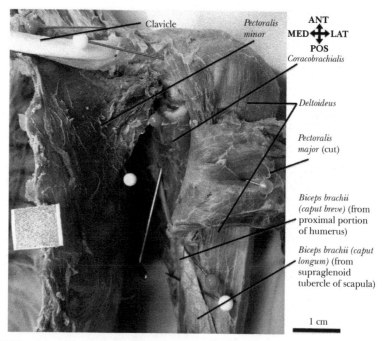

Fig. 20 *Hylobates lar* (HU HL1, adult male): ventral view of the left pectoral and arm musculature after reflecting the pectoralis major.

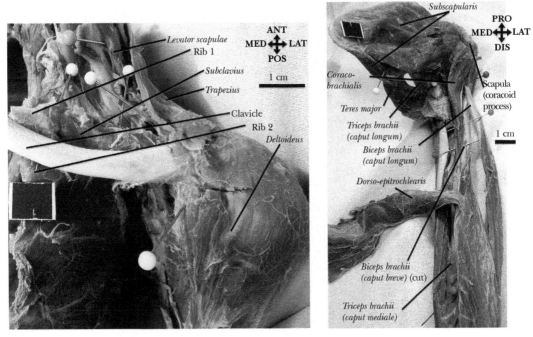

Fig. 21 *Hylobates lar* (HU HL1, adult male): ventral view of the left pectoral and arm musculature, after removal of pectoralis major and pectoralis minor (on the left); ventral view of the left pectoral and arm musculature associated with the humerus (on the right). Note that, as is often the case in hylobatids, in this specimen the short head of the biceps brachii originates from the proximal portion of the humerus (and not from the scapula, as is usually the case in most primates).

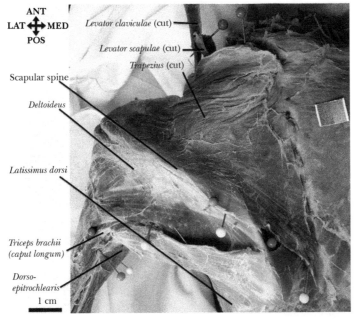

Fig. 22 *Hylobates lar* (HU HL1, adult male): dorsal view of the left pectoral and arm musculature; note that the head was removed.

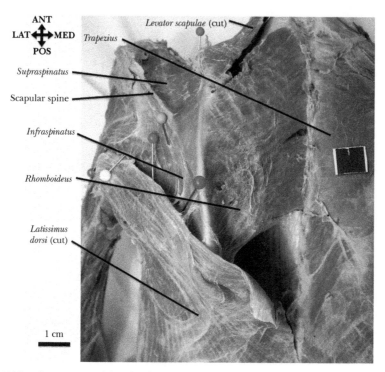

Fig. 23 *Hylobates lar* (HU HL1, adult male): dorsal view of the left pectoral and arm musculature, after removal of trapezius and levator claviculae.

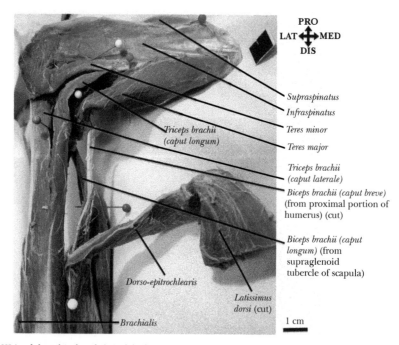

Fig. 24 *Hylobates lar* (HU HL1, adult male): dorsal view of the left pectoral and arm musculature associated with the humerus.

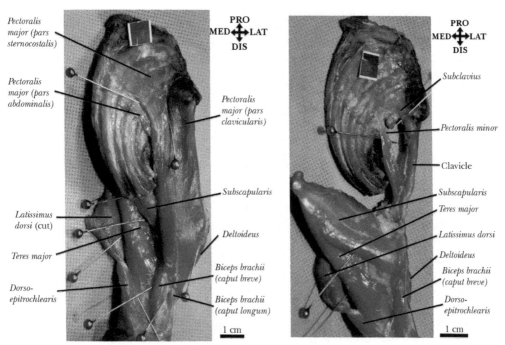

Pectoralis major (pars sternocostalis)

Pectoralis major (pars abdominalis)

Pectoralis major (pars clavicularis)

Latissimus dorsi (cut)

Teres major

Dorso-epitrochlearis

Subscapularis

Deltoideus

Biceps brachii (caput breve)

Biceps brachii (caput longum)

PRO
MED LAT
DIS

1 cm

Subclavius

Pectoralis minor

Clavicle

Subscapularis

Teres major

Latissimus dorsi

Deltoideus

Biceps brachii (caput breve)

Dorso-epitrochlearis

PRO
MED LAT
DIS

1 cm

Fig. 25 *Hylobates gabriellae* (VU HG1, infant male): ventral views of the left pectoral and arm musculature (N.B., on the right picture, the pectoralis major was removed).

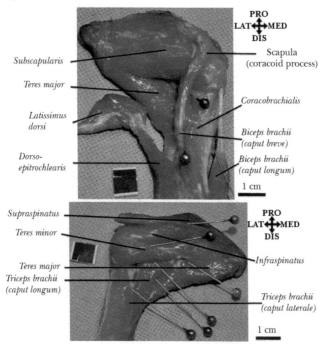

Subscapularis

Teres major

Latissimus dorsi

Dorso-epitrochlearis

Scapula (coracoid process)

Coracobrachialis

Biceps brachii (caput breve)

Biceps brachii (caput longum)

PRO
LAT MED
DIS

1 cm

Supraspinatus

Teres minor

Teres major
Triceps brachii (caput longum)

Infraspinatus

Triceps brachii (caput laterale)

PRO
LAT MED
DIS

1 cm

Fig. 26 *Hylobates gabriellae* (VU HG1, infant male): ventral view of the left pectoral and arm musculature, after removal of pectoralis major, pectoralis minor, subclavius, and deltoideus (on the top); dorsal view of the left pectoral and arm musculature, after removal of deltoideus (on the bottom).

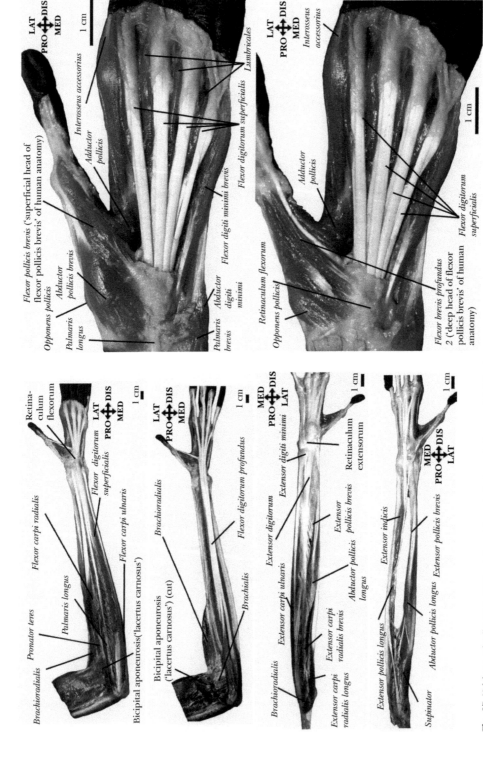

Fig. 27 *Hylobates gabriellae* (VU HG2, adult male): ventral views of the superficial (top) and deeper (second from top) ventral muscles of the left upper limb; dorsal views of the superficial (third from top) and deeper (bottom) dorsal forearm muscles of the left upper limb.

Fig. 28 *Hylobates gabriellae* (VU HG2, adult male): palmar views of the left hand muscles (on the bottom picture the abductor pollicis brevis and flexor pollicis brevis were removed).

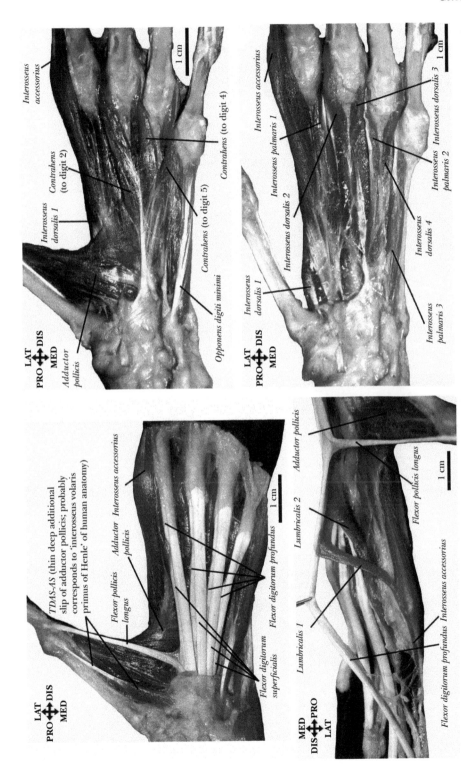

Fig. 29 *Hylobates gabriellae* (VU HG2, adult male): palmar view of the left hand muscles after removal of abductor pollicis brevis, opponens pollicis, flexor pollicis brevis and flexor brevis profundus 2 (on the top); ventrolateral view showing the insertion of the left lumbricales onto the dorsal surface of the flexor digitorum profundus tendons (on the bottom).

Fig. 30 *Hylobates gabriellae* (VU HG2, adult male): palmar views of the deepest muscles of the hand (N.B., on the picture of the bottom the contrahentes, adductor pollicis and opponens digiti minimi were removed).

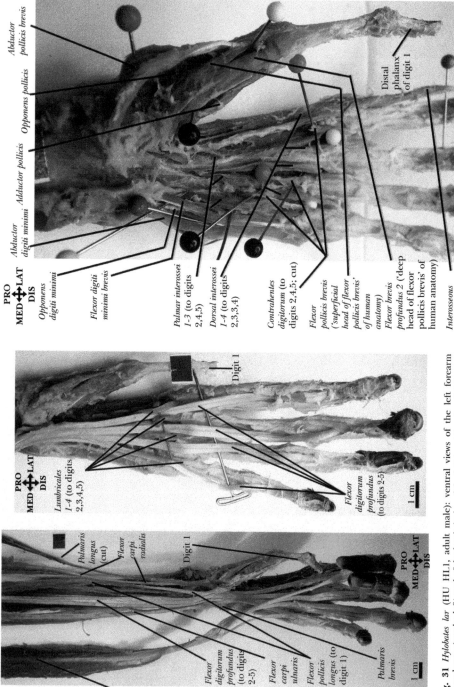

Fig. 32 *Hylobates lar* (HU HL1, adult male): ventral view of the muscles of the left hand; the palmaris brevis was removed.

Fig. 31 *Hylobates lar* (HU HL1, adult male): ventral views of the left forearm musculature (on the left) and of the lumbricales of the left hand (on the right).

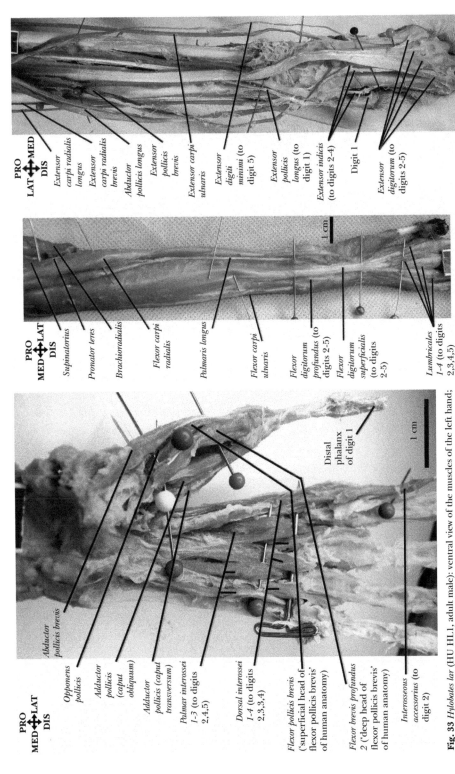

PRO
LAT ← → MED
DIS

Extensor
carpi radialis
longus

Extensor
carpi radialis
brevis

Abductor
pollicis longus

Extensor
pollicis
brevis

Extensor carpi
ulnaris

Extensor
digiti
minimi (to
digit 5)

Extensor
pollicis
longus (to
digit 1)

Extensor indicis
(to digits 2-4)

Digit 1

Extensor
digitorum (to
digits 2-5)

PRO
MED ← → LAT
DIS

Supinatorius

Pronator teres

Brachioradialis

Flexor carpi
radialis

Palmaris longus

Flexor carpi
ulnaris

Flexor
digitorum
profundus (to
digits 2-5)

Flexor
digitorum
superficialis
(to digits
2-5)

Lumbricales
1-4 (to digits
2,3,4,5)

Fig. 34 *Hylobates gabriellae* (VU HG1, infant male): ventral view of the left ventral forearm and hand musculature (on the left); *Hylobates lar* (HU HL1, adult male): dorsal view of the left dorsal forearm musculature (on the right).

PRO
MED ← → LAT
DIS

Abductor
pollicis brevis

Opponens
pollicis

Adductor
pollicis
(caput
obliquum)

Adductor
pollicis (caput
transversum)

Palmar interossei
1-3 (to digits
2,4,5)

Dorsal interossei
1-4 (to digits
2,3,3,4)

Flexor pollicis brevis
('superficial head of
flexor pollicis brevis'
of human anatomy)

Flexor brevis profundus
2 ('deep head of
flexor pollicis brevis'
of human anatomy)

Interosseous
accessorius (to
digit 2)

Distal
phalanx
of digit 1

Fig. 33 *Hylobates lar* (HU HL1, adult male): ventral view of the muscles of the left hand; the palmaris brevis, contrahentes digitorum and hypothenar muscles were removed.

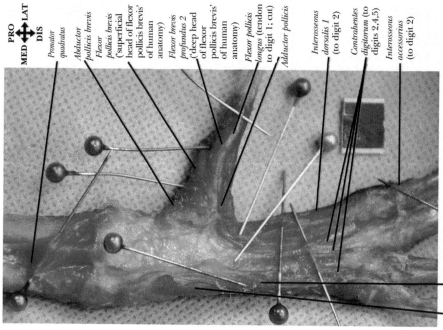

PRO
MED ✛ LAT
DIS

Pronator
quadratus

Abductor
pollicis brevis

Flexor
pollicis brevis
(superficial
head of flexor
pollicis brevis'
of human
anatomy)

Flexor brevis
profundus 2
('deep head
of flexor
pollicis brevis'
of human
anatomy)

Flexor pollicis
longus (tendon
to digit 1; cut)

Adductor pollicis

Interosseous
dorsalis 1
(to digit 2)

Contrahentes
digitorum (to
digits 2,4,5)

Interosseous
accessorius
(to digit 2)

1 cm

Flexor digiti minimi brevis

Abductor digiti minimi

Fig. 36 *Hylobates gabriellae* (VU HG1, infant male): ventral view of the left hand muscles after removal of forearm flexors.

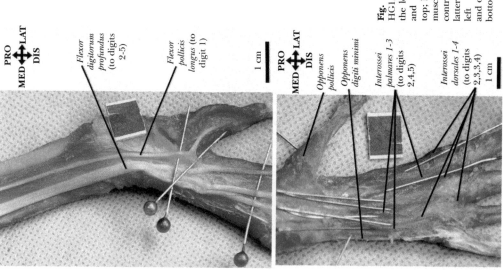

PRO
MED ✛ LAT
DIS

Flexor
digitorum
profundus
(to digits
2-5)

Flexor
pollicis
longus (to
digit 1)

1 cm

PRO
MED ✛ LAT
DIS

Opponens
pollicis

Opponens
digiti minimi

Interossei
palmares 1-3
(to digits
2,4,5)

Interossei
dorsales 1-4
(to digits
2,3,3,4)

1 cm

Fig. 35 *Hylobates gabriellae* (VU HG1, infant male): ventral views of the left flexor digitorum profundus and flexor pollicis longus (on the top; N.B., the tendon of the former muscle to digit 2 also receives a small contribution from the tendon of the latter muscle to digit 1) and of the left interossei, opponens pollicis and opponens digiti minimi (on the bottom).

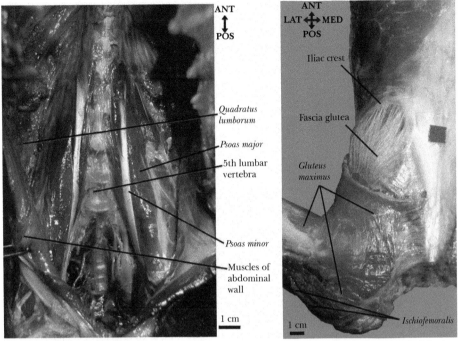

Fig. 37 *Hylobates gabriellae* (VU HG2, adult male): ventral view of the abdominal and pelvic musculature (on the left); dorsal view of the left buttock musculature (on the right).

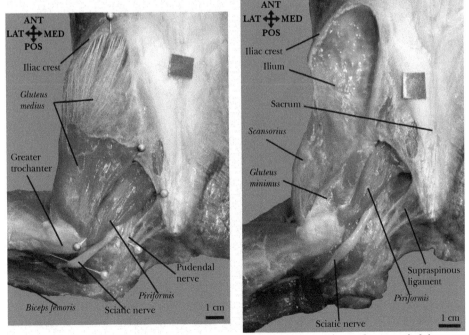

Fig. 38 *Hylobates gabriellae* (VU HG2, adult male): dorsal views of the left buttock musculature after removal of gluteus maximus and ischiofemoralis (on the left) and also of gluteus medius (on the right).

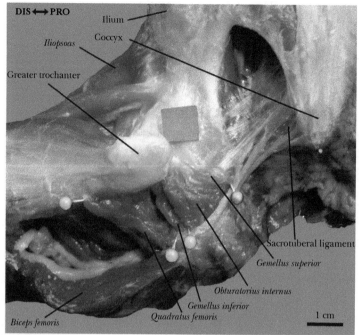

Fig. 39 *Hylobates gabriellae* (VU HG2, adult male): dorsal view of the deep muscles of the left buttock. Note the presence of the gemellus superior.

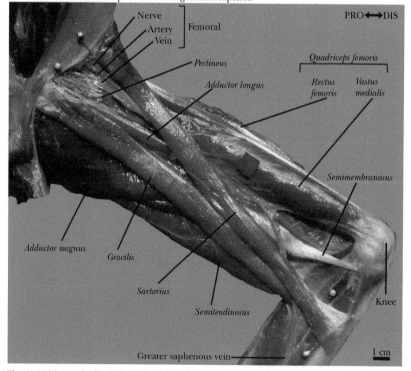

Fig. 40 *Hylobates gabriellae* (VU HG2, adult male): ventral view of the muscles of the left thigh.

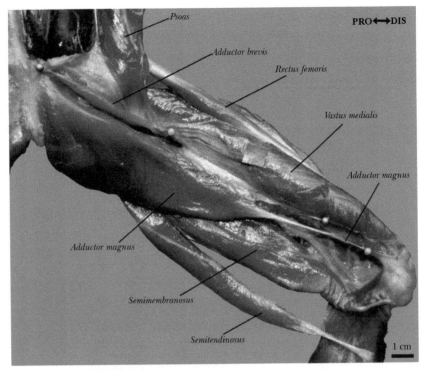

Fig. 41 *Hylobates gabriellae* (VU HG2, adult male): ventral view of the muscles of the left thigh after removal of some superficial muscles. Note that we could not find the muscle adductor minimus.

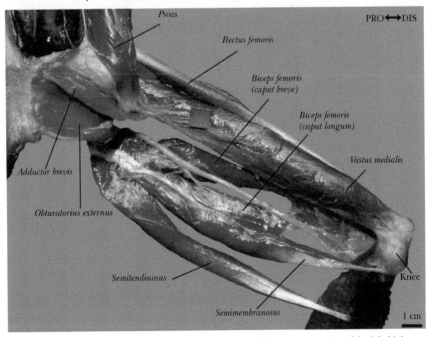

Fig. 42 *Hylobates gabriellae* (VU HG2, adult male): ventral view of the deeper muscles of the left thigh.

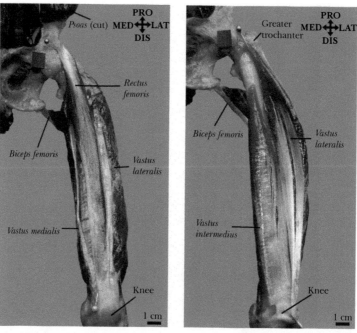

Fig. 43 *Hylobates gabriellae* (VU HG2, adult male): ventromedial views of the muscles of the left thigh (N.B., on the right picture the vastus medialis and psoas were removed).

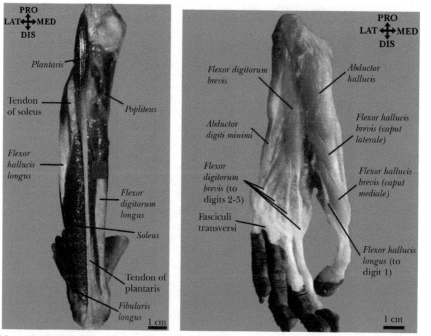

Fig. 44 *Hylobates gabriellae* (VU HG2, adult male): dorsal view of the flexor musculature of the left leg after removal of gastrocnemius (on the left; N.B., the plantaris runs from the tibial lateral condyle to the medial side of the calcaneal tuberosity); intermediate plantar view of the muscles of the left foot (on the right).

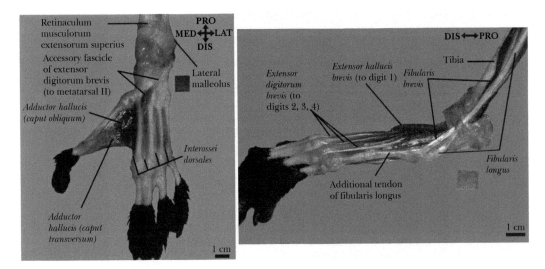

Fig. 45 *Hylobates gabrielae* (VU HG2, adult male): ventral view of the deep musculature of the left foot (on the left; N.B., there is an accessory fascicle of the extensor digitorum brevis to metatarsal II and the first dorsal interosseous has a single origin from the medial side of metatarsal II); lateral view of the musculature of the left leg and foot (N.B., the fibularis tertius is missing and the fibularis longus has an additional tendon: see text).

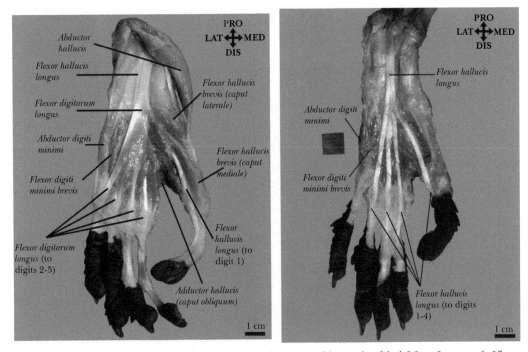

Fig. 46 *Hylobates gabriellae* (VU HG2, adult male): intermediate plantar view of the muscles of the left foot after removal of flexor digitorum brevis and lumbricales (on the left; N.B., the quadratus plantae is missing); deep plantar view of the muscles of the left foot after removal of adductor hallucis (on the right).

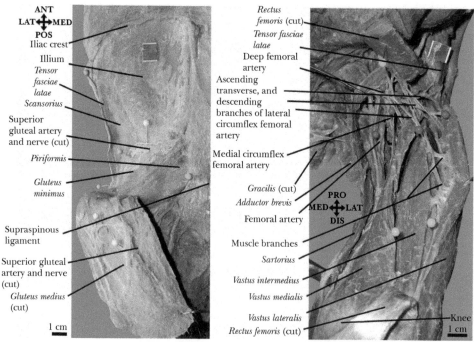

Fig. 47 *Hylobates lar* (HU HL1, adult male): dorsal view of the muscles of the left buttock (on the left; N.B., there is a scansorius lying between the tensor fasciae latae and the gluteus minimus); ventromedial view of the deep muscles of the left thigh (on the right).

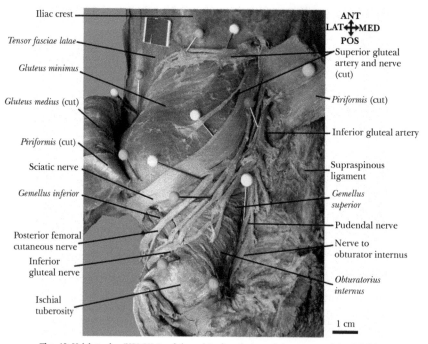

Fig. 48 *Hylobates lar* (HU HL1, adult male): dorsal view of the nerves of the left hip and buttock.

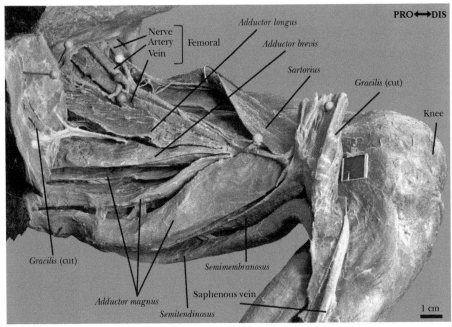

Fig. 49 *Hylobates lar* (HU HL1, adult male): ventral view of the deep muscles of the left thigh.

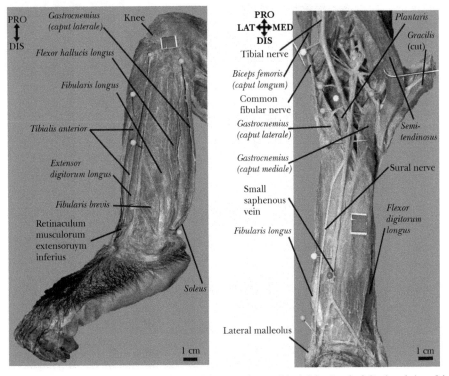

Fig. 50 *Hylobates lar* (HU HL1, adult male): lateral view of the musculature of the left leg (on the left); dorsal view of the flexor musculature of the left leg (on the right).

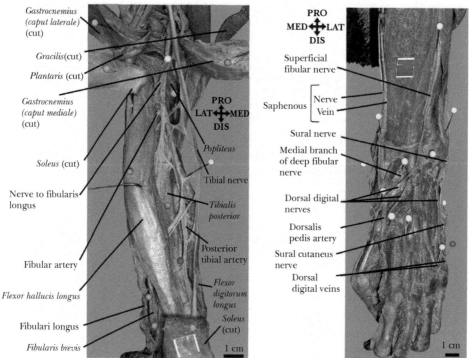

Fig. 51 *Hylobates lar* (HU HL1, adult male): dorsal view of the deepest flexor musculature of the left leg (on the left); ventral view of the superficial nerves and vessels of the left leg (on the right; N.B., there is no fibularis tertius).

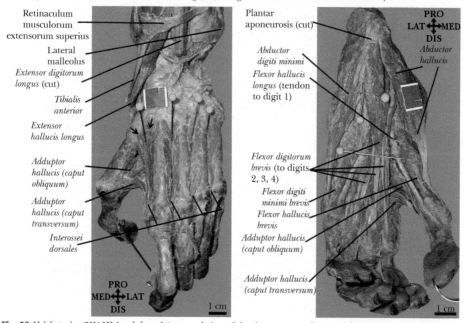

Fig. 52 *Hylobates lar* (HU HL1, adult male): ventral view of the deepest musculature of the left foot (on the left; N.B., the doble head of origin of the first interosseous takes origin from the lateral side of metatarsal I and medial side of metatarsal II—indicated by arrows); superficial plantar view of the muscles of the left foot (on the right).

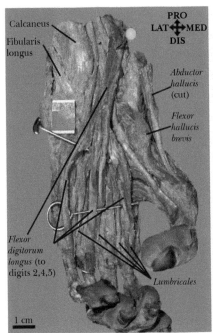
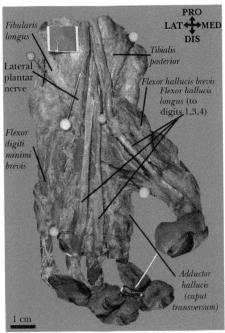

Fig. 53 *Hylobates lar* (HU HL1, adult male): intermediate plantar view of the muscles of the left foot (on the left; N.B. the second lumbrical has a doble origin, the third one is fused with the flexor hallucis longus, and the fourth one is fused with the flexor hallucis longus and flexor digitorum longus: see text); same view after removal of flexor digitorum longus and lumbricales (on the right).

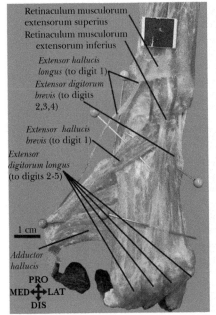
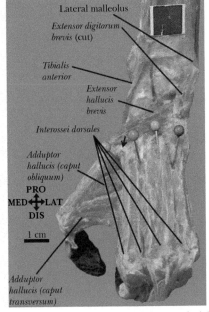

Fig. 54 *Hylobates lar* (GWU HL1, juvenile female): ventral view of the deep musculature of the left foot (on the left; N.B., the tendons of extensor digitorum longus, which insert into the dorsal aponeurosis of digits 2-5, are not well separated; the tendons appeared grouped); ventral view of the deepest musculature of the left foot (on the right; N.B., the single head of origin of the first dorsal interosseous originates from the medial side of metatarsal II—indicated by an arrow).

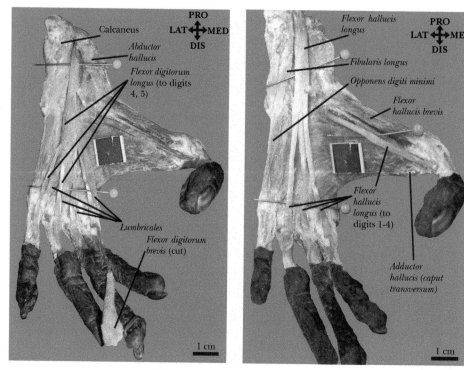

Fig. 55 *Hylobates lar* (GWU HL1, juvenile female): intermediate plantar view of the muscles of the left foot (on the left; N.B., the flexor digitorum brevis peculiarly has a single tendon, to digit 2); intermediate plantar view of the muscles of the left foot after removal of flexor digitorum longus and lumbricales (on the right).

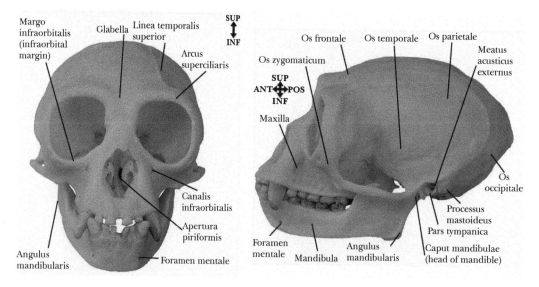

Fig. 56 *Hylobates klossii* (VU HK1, adult male): facial view of the cranium (on the left); lateral view of the left side of the skull (on the right).

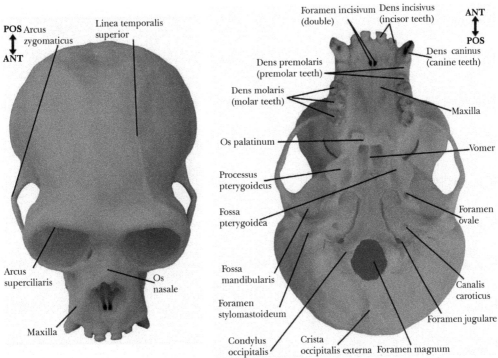

Fig. 57 *Hylobates klossii* (VU HK1, adult male): superior (on the left) and inferior (on the right) views of the cranium.

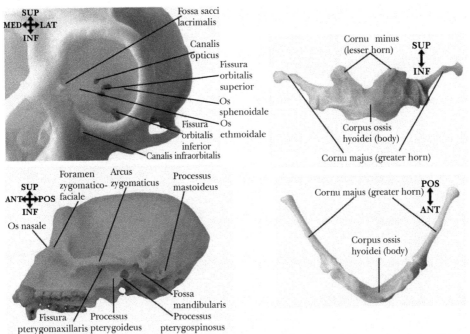

Fig. 58 *Hylobates klossii* (VU HK1, adult male): frontal view of the left orbital cavity (top left); ventrolateral view of the cranium (bottom left); *Nomascus gabriellae* (VU HG1, adult male): anterior (top right) and superior (bottom right) views of the hyoid bone.

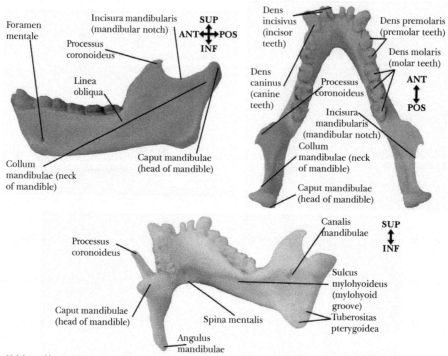

Fig. 59 *Hylobates klossii* (VU HK1, adult male): lateral (top left), superior (top right) and posterior (bottom) views of the mandible.

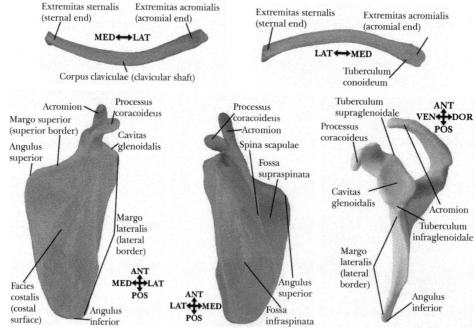

Fig. 60 *Hylobates klossii* (VU HK1, adult male): anterior (top left) and posterior (top right) views of the left clavicle; ventral (bottom left) and dorsal (bottom center) and lateral (bottom right) views of the left scapula.

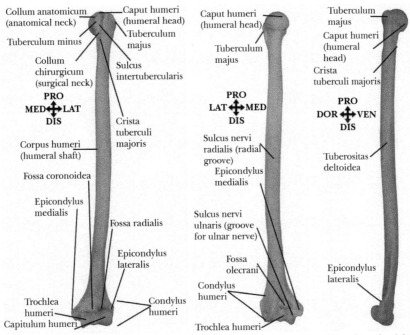

Collum anatomicum (anatomical neck)

Caput humeri (humeral head)

Tuberculum majus

Tuberculum minus

Collum chirurgicum (surgical neck)

Sulcus intertubercularis

PRO
MED ✛ **LAT**
DIS

Corpus humeri (humeral shaft)

Crista tuberculi majoris

Fossa coronoidea

Epicondylus medialis

Fossa radialis

Epicondylus lateralis

Trochlea humeri

Capitulum humeri

Condylus humeri

Caput humeri (humeral head)

Tuberculum majus

Tuberculum majus

Caput humeri (humeral head)

Crista tuberculi majoris

PRO
LAT ✛ **MED**
DIS

PRO
DOR ✛ **VEN**
DIS

Sulcus nervi radialis (radial groove)

Epicondylus medialis

Tuberositas deltoidea

Sulcus nervi ulnaris (groove for ulnar nerve)

Fossa olecrani

Condylus humeri

Epicondylus lateralis

Trochlea humeri

Fig. 61 *Hylobates klossii* (VU HK1, adult male): ventral (on the left), dorsal (on the center) and lateral (on the right) views of the left humerus.

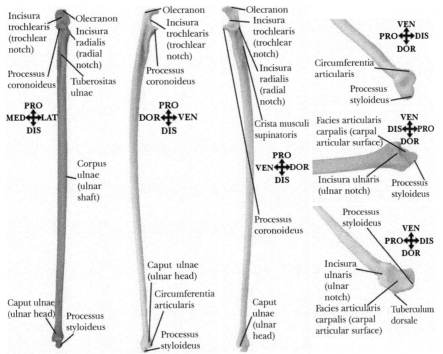

Incisura trochlearis (trochlear notch)

Olecranon

Incisura radialis (radial notch)

Processus coronoideus

Tuberositas ulnae

PRO
MED ✛ **LAT**
DIS

Olecranon

Incisura trochlearis (trochlear notch)

Processus coronoideus

PRO
DOR ✛ **VEN**
DIS

Olecranon

Incisura trochlearis (trochlear notch)

Incisura radialis (radial notch)

Crista musculi supinatoris

PRO
VEN ✛ **DOR**
DIS

Processus coronoideus

VEN
PRO ✛ **DIS**
DOR

Circumferentia articularis

Processus styloideus

Facies articularis carpalis (carpal articular surface)

VEN
DIS ✛ **PRO**
DOR

Incisura ulnaris (ulnar notch)

Processus styloideus

Corpus ulnae (ulnar shaft)

Processus styloideus

Circumferentia articularis

Processus styloideus

VEN
PRO ✛ **DIS**
DOR

Incisura ulnaris (ulnar notch)

Tuberculum dorsale

Caput ulnae (ulnar head)

Processus styloideus

Caput ulnae (ulnar head)

Caput ulnae (ulnar head)

Facies articularis carpalis (carpal articular surface)

Fig. 62 *Hylobates klossii* (VU HK1, adult male): ventral, medial and lateral views of the left ulna (three figures on the left, respectively); details of the distal end of the left ulna (top right) and of the left radius (center right and bottom right).

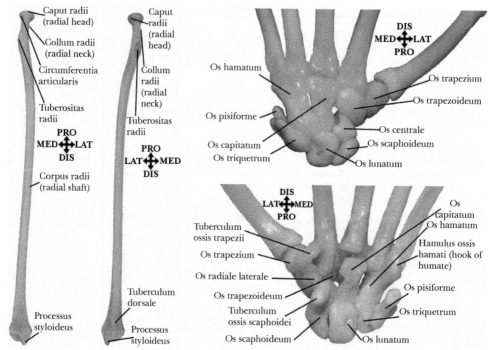

Fig. 63 *Hylobates klossii* (VU HK1, adult male): ventral view and dorsal views of the left radius (two figures on the left, respectively); dorsal (top right) and palmar (bottom right) views of the carpal bones of the left hand.

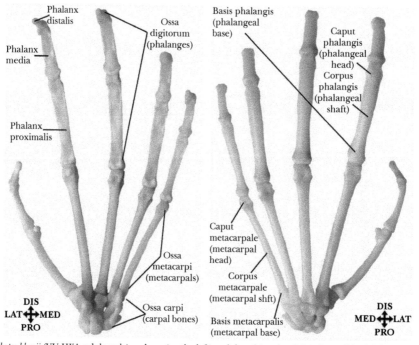

Fig. 64 *Hylobates klossii* (VU HK1, adult male): palmar (on the left) and dorsal (on the right) views of the left hand.

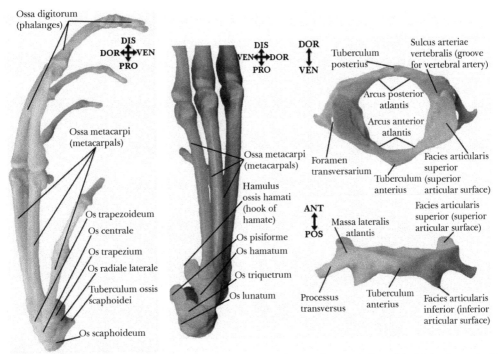

Fig. 65 *Hylobates klossii* (VU HK1, adult male): radial (on the left) and ulnar (on the center) views of the left hand; anterior (top right) and ventral (bottom right) views of the atlas.

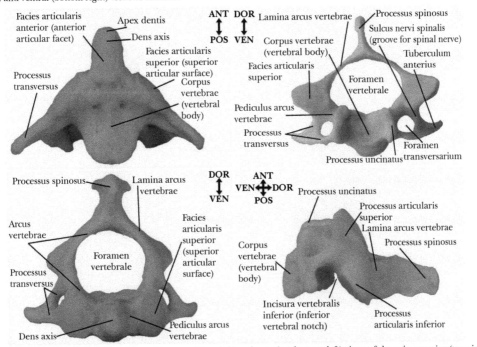

Fig. 66 *Hylobates klossii* (VU HK1, adult male): ventral (top left) and anterior (bottom left) views of the axis; anterior (top right) and lateral (bottom right) views of a typical cervical vertebra.

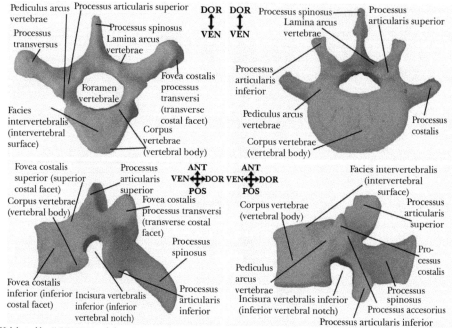

Fig. 67 *Hylobates klossii* (VU HK1, adult male): anterior (top left) and lateral (bottom left) views of a thoracic vertebra; anterior (top right) and lateral (bottom right) views of a lumbar vertebra.

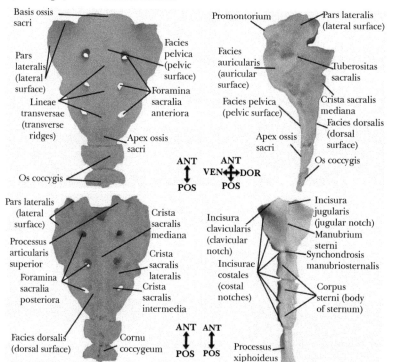

Fig. 68 *Hylobates klossii* (VU HK1, adult male): ventral (top left), dorsal (bottom left) and lateral (top right) views of the sacrum; ventral (bottom right) view of the sternum.

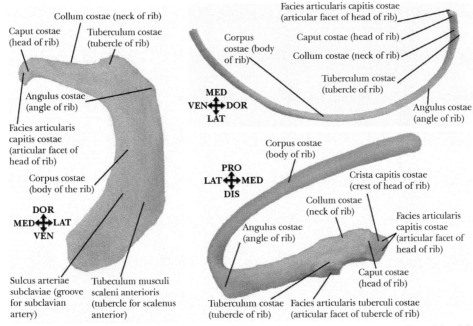

Collum costae (neck of rib)

Caput costae (head of rib)

Tuberculum costae (tubercle of rib)

Facies articularis capitis costae (articular facet of head of rib)

Corpus costae (body of rib)

Caput costae (head of rib)

Collum costae (neck of rib)

Tuberculum costae (tubercle of rib)

Angulus costae (angle of rib)

MED VEN DOR LAT

Angulus costae (angle of rib)

Facies articularis capitis costae (articular facet of head of rib)

Corpus costae (body of the rib)

Corpus costae (body of rib)

PRO LAT MED DIS

Crista capitis costae (crest of head of rib)

Collum costae (neck of rib)

Facies articularis capitis costae (articular facet of head of rib)

DOR MED LAT VEN

Angulus costae (angle of rib)

Caput costae (head of rib)

Sulcus arteriae subclaviae (groove for subclavian artery)

Tubeculum musculi scaleni anterioris (tubercle for scalenus anterior)

Tuberculum costae (tubercle of rib)

Facies articularis tuberculi costae (articular facet of tubercle of rib)

Fig. 69 *Hylobates klossii* (VU HK1, adult male): anterior view of the first left rib (on the left); anterior (top right) and dorsal (bottom right) views of the seventh left rib.

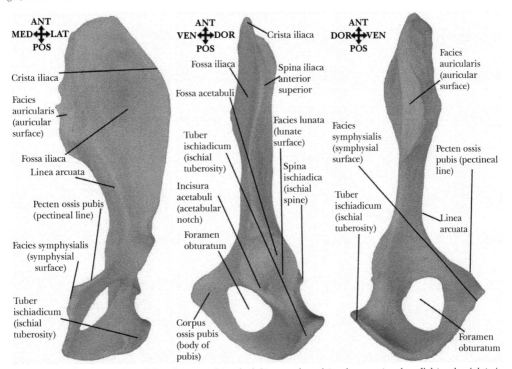

ANT MED LAT POS

Crista iliaca

Facies auricularis (auricular surface)

Fossa iliaca

Linea arcuata

Pecten ossis pubis (pectineal line)

Facies symphysialis (symphysial surface)

Tuber ischiadicum (ischial tuberosity)

ANT VEN DOR POS

Crista iliaca

Fossa iliaca

Fossa acetabuli

Spina iliaca anterior superior

Facies lunata (lunate surface)

Tuber ischiadicum (ischial tuberosity)

Incisura acetabuli (acetabular notch)

Spina ischiadica (ischial spine)

Foramen obturatum

Corpus ossis pubis (body of pubis)

ANT DOR VEN POS

Facies auricularis (auricular surface)

Facies symphysialis (symphysial surface)

Tuber ischiadicum (ischial tuberosity)

Pecten ossis pubis (pectineal line)

Linea arcuata

Foramen obturatum

Fig. 70 *Hylobates klossii* (VU HK1, adult male): ventral (on the left), ventrolateral (on the center) and medial (on the right) views of the left pelvic bone.

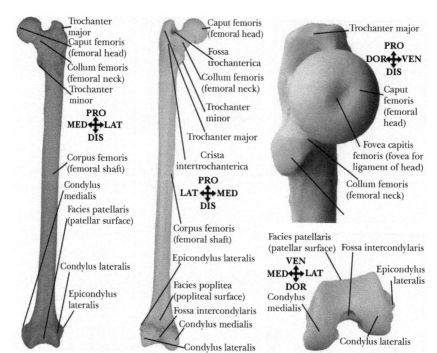

Fig. 71 *Hylobates klossii* (VU HK1, adult male): ventral (on the left) and dorsal (on the center) views of the left femur; medial view of the proximal end (top right) and axial view of the distal end (bottom right) of the left femur.

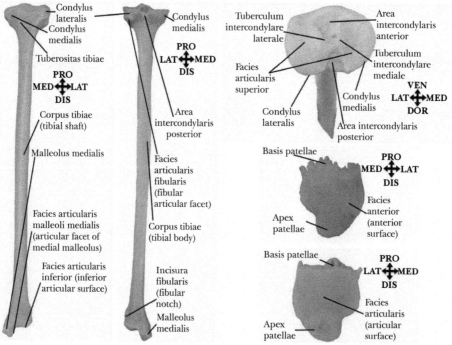

Fig. 72 *Hylobates klossii* (VU HK1, adult male): ventral (on the left) and dorsal (on the center) views of the left tibia; axial view of the proximal end of the left tibia (top right); ventral (center right) and dorsal (bottom right) views of the left patella.

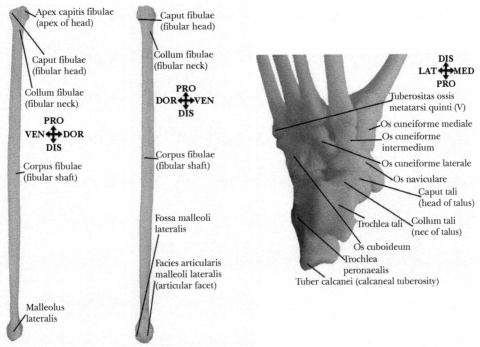

Fig. 73 *Hylobates klossii* (VU HK1, adult male): lateral (on the left) and medial (on the center) views of the left fibula; dorsal view of the tarsal bones of the left foot (on the right).

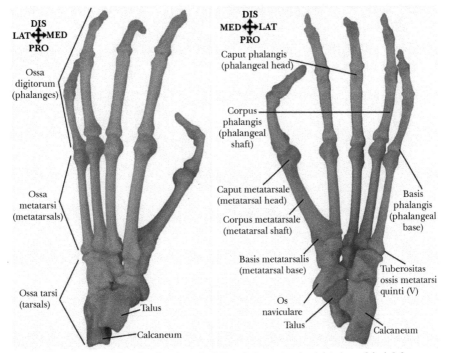

Fig. 74 *Hylobates klossii* (VU HK1, adult male): dorsal (on the left) and plantar (on the right) views of the left foot.

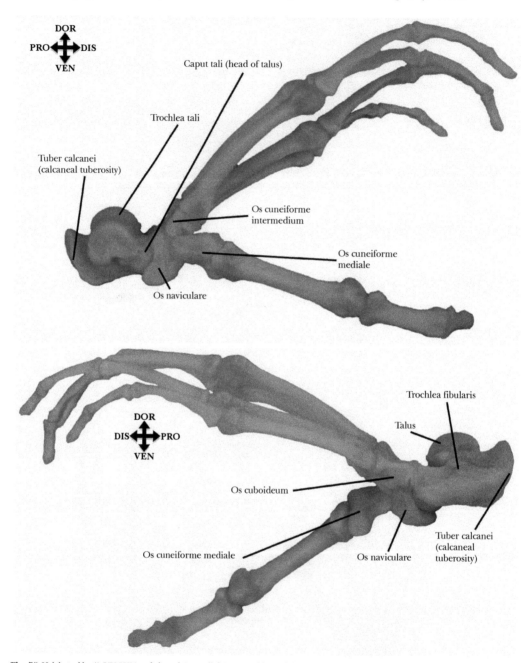

Fig. 75 *Hylobates klossii* (VU HK1, adult male): medial (top) and lateral (bottom) views of the left foot.

Milton Keynes UK
Ingram Content Group UK Ltd.
UKHW050451071024
449327UK00015B/329

9 780367 381509